PARASITOLOGY

NIPA® GENX ELECTRONIC RESOURCES & SOLUTIONS P. LTD.

New Delhi-110 034

About the Author

Dr Mahesh Chandra Agrawal was born on 26th April 1944 in a business family at Firozabad, Uttar Pradesh and was reluctant to join family business and desired to choose academic life. He joined College of Veterinary Science and Animal Husbandry, Jabalpur in 1963, was best student of the college and passed BVSc & AH, MVSc with university gold medals, being first in order of merit in Jawaharlal Nehru Agriculture University. He was awarded ICAR Senior Research fellowship and completed his Ph.D. in Parasitology in 1978. With a short tenure of field services, he joined Jawaharlal Nehru Agriculture University, Jabalpur Veterinary College on 15th January, 1968 as demonstrator, promoted to different posts and retired as Dean of Jabalpur Veterinary College in January, 2005. After retirement, he was bestowed ICAR emeritus scientist award after ten years in the Veterinary faculty to work on control of schistosomiasis under field conditions.

He has a vast experience of teaching undergraduates and guiding post graduate students, both in Veterinary faculty and Zoology/life sciences, faculty of Rani Durgavati Vishwa Vidyalaya .Within limited facilities, he innovated many new technologies like maintenance of fresh water snails in laboratory, infecting large animals with schistosome cercariae, use of vertical water pump for perfusion of large animals for recovering blood-flukes etc. He also advanced science by demonstrating *Bivitellobilharzia nairi* in a new nidus, demonstration of eggs in *S.incognitum-mouse* model, hypoglycemia in schistosomiasis, under estimation of schistosomiasis in south Asia, lower efficacy of praziquantal against Indian schistosomes etc.

Dr Mahesh Chandra Agrawal received many awards including ICAR National fellow which happened to be the first in the field of Parasitology in India he was elected fellow of Zoology Society of India, Indian Association for Advancement of Veterinary Parasitology, National Academy of Veterinary Sciences, acted as Chairman/co-chairman/ Rapporteur in national scientific conferences on Zoology, Veterinary Science, Parasitology and became judge for young scientist award, best research paper etc and has published more than 200 research papers, review articles, and chapters in scientific books.

Beside the present book, he has published following other books

1. Schistosomes and schistosomiasis in South Asia. Year 2012, Springer, India, Page 251, ISBN 9788132205388
2. Remembering Dr S C Dutt: The Parasitologist (Editor). Year 2015, Har Anand Publications Pvt Ltd, New Delhi, page 208, ISBN 9788124119174
3. Ye hai hamare krishi vishwa vidyalaya. Year 2017. Blue Rose Publishers, New Delhi, Page 170. ISBN 9789386673374

Those interested, may follow below blogs

1. www.indianschistosomiasis.blogspot.com
2. www.indianparasitologists.blogspot.com
3. Jvc Alumni Association.blogspot.com

PARASITOLOGY
A Research Guide

M.C. Agrawal

BVSc & AH (Gold Medalist), MVSc (Gold Medalist), Ph.D, FZSI, FIAAVP,
FNAVS Former ICAR National Fellow, Emeritus Scientist
Professor and Head, Department of Parasitology
Dean College of Veterinary Science & AH
Nanaji Deshmukh Veterinary Science University
Indira Gandhi Marg, Panagar, South Civil Lines, Jabalpur
Madhya Pradesh-482001

NIPA® GENX ELECTRONIC RESOURCES & SOLUTIONS P. LTD.
New Delhi-110 034

NIPA® GENX ELECTRONIC RESOURCES & SOLUTIONS P. LTD.

101,103, Vikas Surya Plaza, CU Block
L.S.C. Market, Pitam Pura, New Delhi-110 034
Ph : +91 11 27341616, 27341717, 27341718
E-mail: newindiapublishingagency@gmail.com
web: www.nipabooks.com

For customer assistance, please contact
Phone: + 91-11-27 34 17 17
Fax: + 91-11- 27 34 16 16
E-Mail: feedbacks@nipabooks.com

ISBN: 978-81-19235-91-9

Composed and Designed by NIPA®.

Preface

While I was writing the book "Schistosomes and Schistosomiasis in South Asia" (Published by Springer in June, 2012-ISBN 9788132205388) it struck to me that this reference book is for senior researchers, but what's about our young aspiring scholars who are not quite familiar with scientific research though desiring to take up it as their career. This is because I witnessed difficulties of our post graduate students while pursuing their research problem; how they are ignorant about basic research principles and their wandering here and there to solve their minor problems. Even our Bharat Ratan scientist Dr CNR Rao in his autobiographical book "Climbing the Limitless Ladder - A Life in Chemistry" has mentioned "I had a romantic notion about doing research as a scientist, but had nobody to guide me" He was wondering which college will be best for doing MSc in chemistry for making research as a career (finally he selected BHU) and sadly the state of affairs has not changed much even when we have entered in twenty first century. Though there are many books to guide our young generation to become doctors or engineers or even entrepreneurs, yet there is dearth of books to guide them how and why to choose research as a career and pursue research as goal of their life. Perhaps, this is one reason why Indian universities have failed to attract talent in research thereby remaining laggards in world ranking where research is the main criterion to judge superiority of any institution over others.

Recently ICAR and UGC have provided two courses for postgraduate students; one is about consulting library i.e. scientific literature while other course is for writing a research thesis. Nevertheless, there are no books which can deal the subject comprehensively and teachers are still depending on their old class notes. Even a newly posted Assistant Professor faces great difficulty in formulating a research scheme or publishing research papers, particularly when there is no senior faculty member in the department.

However, it's of limited value to tell the aspiring students only about consulting the literature, or writing and publishing a research paper. Rather, it will be more advantageous to expose them to other aspects of research and to impress that research is essential for advancement of human civilization. In that sense, this book is a treatise to inspire our young generation to join research as their career; this book tells the scholars about benefits of joining research as a career with characteristics desired while joining research as a career.

To deal the subject, I have taken examples from Parasitology - a field of my expertise and a part of life sciences where we study Parasites or creatures which are living on other living organism and causing some harm by their life style.

When I say Research in Parasitology, I do not mean Parasitology in its restricted sense of treating parasitic diseases in man and animals . The place of the subject in life sciences has rightly been stated long back by Professor AC Chandler (1946) "Parasitology touches upon or overlaps as many other sciences that a Parasitologist probably has to stick his nose into more different fields of knowledge than any other kind of biologist. Indeed, modern parasitology might well be designated as 'symbiology' the one of the first truly inter-disciplinary biological sciences". Indeed, in twenty first century, it extends its tentacles that encroaches not only all branches of life sciences (immunology, pathology, biochemistry, microbiology, pharmacology, nutrition , zoonosis, public health , genetics, zoology, botany, malacology, entomology, paleontology, taxonomy, forestry) but also distant subjects such as geology, meteorology, environmental sciences, distance imagination science, hydrology, mathematics, history, economics and sociology.

With my more than forty years experience in scientific research and witnessing the problems being faced by our young scientists while pursuing their career in research, I have tried to cover almost all these problems in this book. It is also to inspire our grand children Kush, Tanav, Ruhan and many others like them to make research as their career in life sciences which is linked to many disciplines of science and humanities.

This is the book which will provide you some details how you may join research activities in Parasitology and carry over research work in your career.

I wish to extend my heartiest thanks to Shri Utsav Agrawal and Shipra Agrawal for extending all the helps. My thanks are also to M/S New India Publishing Agency, New Delhi for taking trouble in editing and publishing this book.

Jabalpur **Mahesh Chandra Agrawal**

26th December, 2017

Contents

Part C

Part D

PART-A

1

Why Interested in Research

If you are inclined to make research as a career, the first question is why are you interested in research or making it as a career in your life - whether for money or to pass a cozy life in a government institute or to get fame or to solve some scientific problems - related to cosmos, Agriculture, animals, human health, search some medicinal plants or you have great passion for science and impressed with its power to find solutions for many universal problems ? Or it is un-employment, now rampant in every field, which is deriving you to join higher degrees and research. Many times, this unemployment or no vacancies in due time leads the young persons to continue their post graduation or joining doctorate program with the hope that some vacancy will arise in the ensuing period or a suitable job may fall after or prior getting a Ph.D. degree.

Searching answer of the above question is linked with another question – what is the aim of your life ? Or what is the dream of your life ? Whether this is your own or affected from outside influences ? Mostly, this is motivated by the image of a profession in the society. Not long ago, when the country was fighting against colonial raj, there was a good social image of *Vakeel* or lawyer; this was since most of our stalwarts, Motilal Nehru, Jawaharlal Nehru, Mahatma Gandhi, were qualified lawyers. As they were icons of that era, most of our young generation followed their path. This craze to become a lawyer faded with the freedom of the nation.

As the newly independent nation was to build roads, power plants, dams, hospitals etc, for welfare of its population, a new demand of engineers and doctors came up and our young generation diverted their attention to choose science subjects for becoming doctors and engineers. This early generation was having a clear aim of their life hence were able to build world class dams, roads or hospitals with bare minimum facilities and earned name and fame for their devotion. However, this devotion started vanishing and in late eighties, when we asked aim of our young generation, their aim was to become an engineer or a doctor. Perhaps, this new generation became confused in searching its aims. Becoming a doctor or engineer may be the means to achieve some goal like serving sick people in the village, opening a world class facility for heart surgery, or constructing a dam to solve water problem etc but if a student considers means as the goal, it will prove a disappointing decision of his life and he may repent later for leaving another lucrative field of work. Most probably becoming a doctor or engineer were not their passion but were choosing these professions by going with the masses or may be by compulsions from their families. This is the reason why many doctors and engineers shift their profession in later part of their life ; and many female applicants do not advance their profession. Therefore it is imperative of deciding aims or profession as per your choice for getting satisfaction in ones' life. If you are selecting the profession as per your passion, and interest, you will certainly achieve important positions since the work will not tire you ; all you will do will look as if you are fulfilling your hobbies.

You will notice that a good proportion of our young generation wants to be engineer, doctor, chartered accountant, lawyer, entrepreneur (now a days, financial manager or MBA) but there are hardly any number of students who are interested in research as their first choice. This may be because our young generation is quite familiar with above professions but ignorant about a research job or have developed a stereotype image of a scientist who is totally submerged in his research without caring for his food or cloths or family. This is far from the truth. Research as a career will be like any other profession and is equally demanding like any other creative profession. Its true that our students or the society is not exposed to the basics of research and is ignorant about many facets of research- perhaps due to absence of the books in this regard. For this reason, once a person is involved accidently in research, it changes his life, he loves to investigate some thing new not only in his subject but every where. His psychology changes towards the whole world and finding a new theorem or answer of a natural phenomenon makes him a special person not only for his society but internationally- not only to his generation but for future generations.

There are many examples of famed persons who devoted their life in searching answer of one or other problem but prior their discovery, they were like an ordinary person. Falling an apple in the garden, compelled Sir Isaac Newton to think why an apple, or for that matter any subject, comes down only to earth. This resulted in discovery of law of gravity. Or Edward Jenner noted absence of small pox or low severity of the pox in his milk maid who had close association with her milch cow. This resulted in an innovative method of vaccinating the persons against small pox and later against many pathogens.

There have been some crazy persons who have taken research as aim of their life and devoted whole life for finding the truth -without caring awards or recognitions or bitter criticism. Perhaps, you may not be aware that Edward Darwin had to face severe criticism from Catholic Churches for his hypothesis of 'origin of species' where he disputed that God has created different animals, present on the earth. This tells us a bitter truth that true research is neither simple nor always provides you elevations or laurels. Contrarily, it may land you in trouble. How one can forget the death penalty extended to Galileo by his claiming that it is earth which is rotating round the sun and not vice versa. Neither it is necessary that your fundamental research will bring always laurels to you. There is the story of Russian botanist Nikolai Vavilov who had influenced with the discovery of Mendel's principles of genetics, and was searching wild ancestors of domestic crops, during 1920-30 in Russia so that these crops may be improved genetically for more food production. But the work of Vavilov was not liked by Joseph Stalin, Head of the then Soviet Union, as he was under influence of less capable botanist Trofim Lysenko whose main thesis was that it is the climate and not genetics which dictated plant development. Though Vavilov was correct (as latter works proved) but he was arrested and starved to death in a Soviet jail (1). Thus the great botanist met with starvation and death in Soviet Union to remain stick to his just thesis while the other enjoyed state patronage on the basis of thesis which proved wrong.

In twenty first century, times have changed a lot and all the credit for this must go to our iconic scientists particularly Einstein for discovery of his atomic theory resulting in atomic exploration by several countries; nobody now can imagine above harassments, that too in a democratic world. However, a somewhat similar situation may arise when your research work disproves the hypothesis of your peers or heads of the institutions- they may try to suppress your research publication or at the most may harm your career. No doubt, the situation becomes problematic but remember that facts or truth always remain true even if someone try either to suppress or falsify it. However, it must be accepted that these are extreme cases which are witnessed rarely.

Nevertheless, every country has now realized importance of research in every field, thanks to our contemporary scientists; it may be in Agriculture, health, automobiles, astrophysics, or any other field. This is also because America has attained super-power status surpassing one time super power Britain, only because of innovations and research achievements which brought not only highest number of noble prizes to that country but also development of new technologies which are important in every sphere of life. Therefore, joining a research field , in present days, is providing recognition as well as money. If you are talented, join research and fulfill your dream by achieving your life time goals. Research as a career gives you an opportunity to do some unforgettable work which may be remembered centuries, afterwards.

It is better to know more about scope and problems, in present day scenario, if you select research as a career. It is emphasized that research is a most creative profession like any other creative profession hence the achievements will solely differ from person to person-so is happening in literature, art and film. Outcome of research work is a cumulative result of your personality, curiosity, eagerness to know, intelligence, your sincerity of investigations and hard work which you have done. Therefore, if you are joining research career due to unemployment or without any interest to know the unknown, there are all chances that you may repent in later part of your life- as neither you will achieve any outstanding result nor will feel financially satisfied. Ironically, I have witnessed some of my colleagues to repent not to join a post of Police Inspector or Revenue officer, as these posts, in their opinion, draw more power and money while research or teaching is a dull profession without any thrill. And in their eyes, the post of Police inspector was greater than a university teacher. Some scholars prefer a research post merely to pass a cozy life–sitting in air-conditioned room and repeating foreign research work so that their bio-data may boost research publications in their account. It's difficult to imagine if these persons feel satisfaction, after their retirement.

If you plan to make research as your life time goal, it's better to know a little bit about modern scientific research which may be classified in two main groups. One is Biological sciences which includes research of all living creatures including plants. And other is Physical sciences which deals with non-living matter of the universe. With advancement of science, many new branches of scientific research have evolved which combine knowledge of these two branches of science. The best example is Biochemistry where chemistry is combined with biology. Similarly, a more new branch is biophysics and bio engineering where physics laws are tried to study biological phenomenon. Likewise, mathematics is being used to develop mathematical models for epidemiological studies- these mathematical formulae are helpful in deciding minimum thresh hold level of a pathogen required for its spread in a given population. Statistical analysis is an important tool in any research work to conclude whether the results are significantly varied or are non-significant.

Beside culmination of two disciplines, physical sciences have been used to develop X-Ray units, CAT scanning or Ultra sonography which are important for diagnosing particular ailment. Therefore, it is not necessary that one has to select either biological sciences or physical sciences, rather it is individual interest which will decide his field of research work. In fact, life is very short to deal any field of research and work of an imminent scientist of his whole life can be summed up as a drop in an ocean- almost all the scientists have devoted their whole life for solving only one vital problem.

This book deals with research in Parasitology - a field of life sciences where we study about the Parasites or creatures which are living on other living organism and causing some harm by their nature of living. These parasites belong to two main biological groups- one is plant kingdom which incorporate bacteria, virus, fungus and other is animal kingdom which incorporate protozoa, helminths, arthropods and Parasitology studies those which belong to Animal kingdom.

Parasitology opens up the ways to study how and why one organism, instead of living freely in its habitats, search another organism to live upon i.e. it is to find shelter and uninterrupted food-the two most important conditions for survival of any creature on earth; how even another parasite treated this parasite as its host making the event as hyper-parasitism; what difficulties these parasites observed in its new host or new environment and how the parasite survived within the host; what are the mechanisms of searching a suitable host by the parasite; how a parasite, devoid of eye, nose and ear, search its partner for mating; how the parasite continued its evolution in synchronization with its host; how it travelled different continents with the migration of its host; what mechanisms host devised to get rid of the parasites; how drugs have been evolved to kill the parasite; what is the pharmo-kinetics and pharmacology of the parasiticides; why we have not been able to develop suitable vaccine against any parasite like malaria, trypanosomes; what are the mechanisms of biological phenomena observed in the life of a parasite like metamorphosis, eccdysis, excystation of infective stage; what are the trigger mechanisms for each of these events.

Study of Parasitology enlarges when we know that the parasite are thousands in number and do not belong to one biological group but varied groups like helminthes (again flukes, tapeworms, round worms each group again classified in many), Protozoa (again of many types like mastigophora, Sarcodina, Apicomplexa), large number of insects (louse, flea, flies), ticks (hard and soft ticks) and mites (more details in chapter 8). These parasites use almost all the organisms as their hosts- in some hosts they survive without much damaging the host while in others it becomes lethal. Almost all the invertebrates like snails, flies, beetles, bugs, mosquitoes, are acting as intermediate host (where parasite develops immature or asexual stages) for one or other parasite while all the vertebrates like mammals, birds, reptiles, fishes are the host for one or other parasite-some of them might be less harmful while others may be lethal- its interesting to study reasons for these differences.

It is better to understand how vast a tiny subject becomes when we start investigating it. For instance, Bilharziasis or schistosomiasis is an ancient disease, occurring both in man and his animals in most of Asian and African countries. The disease leads to hematuria or passing of blood in urine or diarrhea and dysentery in man while animals show either diarrhea or snoring due to nasal granuloma. The etiological agent is a fluke (*Schistosoma*) which passes its life cycle both in mammals and in a fresh water snail. There are various dimensions in studying schistosomiasis (2). It includes studies on the schistosomes and their species (Helminthology, taxonomy, electron microscopy), survival of the parasite in the nature (epidemiology), studies on the fresh water snail (Malacology), harms caused to man, animals, snails (pathology, immunology), methods of identifying the infection (parasitological diagnosis, immunodiagnosis), its treatment (chemotherapy, pharmacology) and prevention of the infection. Again, some new fields of research have opened where scientists are working on the phylogeny (genetic studies) of schistosomes, evolution of schistosome species and mode of its spread to different countries, zoo-geology, effect of climate and ecology responsible for sustenance of the parasite and its snail, genetic make up of human race making the particular race prone or resistant to the infection. To this we may add mathematical models which are worked out to know threshold number of infected cases to sustain the infection in particular environment and economical and sociological events that happened due to schistosomiasis.

This book is indented to such devoted talented persons who have courage to work under adverse circumstances and make remarkable contributions towards human life. It may be considered a biased statement but its true that while physical sciences or engineering discipline is destined to work for comfort of human beings, research in parasitic diseases or Parasitology helps in solving miseries of humans who are suffering from different diseases or their animals - the cure of animals results not only help to voiceless creatures but also enhances economies of the poorest of poor persons present any where in the world.

Therefore, it is evident that you may join any branch of science to study the subject of parasites and parasitology provided you are interested in doing research. If you are passionate about scientific research and ready to make research as your career, this book is written for you to achieve your goals in scientific research.

2

Historical Events

Research is the basics which have changed human life and surrounding environment what existed on earth million years ago into present form. As soon as man realized himself as a separate entity from other animals, he started thinking about himself, creatures co-inhabiting with him and the universe. In that sense, biological science is the oldest subject which fascinated man, though it was called natural sciences in ancient times. Perhaps, the first curiosity was about himself, his birth, death and alike matters. As soon as he realized mortality of life or faced death of his dear ones, he attempted to become immortal or *'amar'*. This is reflected by worshiping of *'Asur '*(*Hirnakashyap, Bhasmasur, Ravan* etc) to *'Brahma'* or *'Shiva'* for receiving the blessings of becoming immortal . However, even God refrained to bless these personalities becoming *'amar'* confirming the realization by all that life is strongly linked with death ; as such they were blessed with certain conditional ties which may lead their death. Eventually, they met with their death under specific circumstances (3). When God or His little part (*Ansh*) took birth on earth as incarnation, He also followed all the rules and regulations of cosmic laws. But He never tried to become immortal and left His body on the earth, after finishing His duties.

There are two facts which we may extract from these mythological stories. First is that there are some cosmic laws which regulate whole universe and no one can escape or break these laws- though these may be used for our benefit. All the scientific research has been directed for understanding these cosmic laws and employ them for human welfare. Second is that man was fascinated with the idea of being 'immortal' since beginning of his journey. This attempt of being *'amar'* diverted human activities in many directions. The most important are the ways and means which man tried to enhance longevity of his life which later led to development of health sciences namely Ayurveda (4). Curious enough it may sound that many years later, the man developed the theory of being *'amar'* by dividing himself into two; one is his body which is perishable and the other is *'atma'* or soul which is immortal. This theory is well reflected in "Shrimad Bhagavat Geeta"(5). Today, some Indian philosophers justify the theory by considering *'atma'* as DNA or molecules which remain unbreakable even after death of any organism. Perhaps, the philosophy of 're-birth' in Hindu religion is a beautiful extension to remain alive for ever in one or other form- how beautifully man has fulfilled his desire of being immortal (5).

In these early attempts, man was trying to understand life and death. The journey of human civilization has witnessed different theories of life and these are reflected in various stories passed from one generation to other. It took long time to understand life and different schools of thought existed to explain life on the earth. Earlier, the theme was that life can evolve from 'nothing' (this was due to inability to observe micro things like small eggs of the insects). Even the great Greek philosophers Hippocrates and Aristotle (350 BC) believed in spontaneous generation of life (is it not surprising that man did not pay attention that he is reproduced only from a mother-a living creature and not by a non-living entity or perhaps he considered himself different from other creatures and thought that other creatures can be created spontaneously). stories of appearing a creature, all of a sudden, may perhaps reflect the same ; reproductive process was under a mystery - even drinking of semen was considered giving birth to a offspring; there was confusion about change of sex or morphology ; a strict demarcation of male or female by birth was not believed as is reflected by changing body structure or sex as per will ; there are many stories of change of man to animal and vice versa in our mythological books reflecting no fixed rules for such phenomena. Nevertheless, the continued and sincere studies of nature confirmed that life is originated only from life and one species cannot transform to another on its will ; sex is determined in mammals only at time of birth with no possibility of its change, afterwards. Another beautiful philosophy propagated in Hinduism was that all creatures are equal and have an identical soul-this timeless theory has proved correct by different scientific discoveries proving an identical cell (or DNA)is the basic unit of life in all creatures existing on the earth.

From the time of origin of man as *Homo erectus* and then *Homo sapiens* on earth, about 10,00,000 or 2,00,000 years ago, he spend almost all his time, except last few thousand years, not much different from other animals. The first important change was walking on two hind legs and using fore legs for holding and using stones, bones as tools; both were distinctive changes, never seen in any other animal species. This along with the use of fire and wheel (non-biological discoveries) provided man special powers by which he was able to control other animal species and to spend a different life style. Then occurred three important biological events which changed human life completely. The first was the important invention of Agriculture (6) (latin Agricultura- ager = field, cultura = cultivation), a most important biological research, occurred about ten thousand years BC; prior that man was hunter-gatherer and was passing almost all his time in searching food and saving it for crisis period. However, by Agriculture, Man learnt to grow crops by burrowing seeds in the soil and getting the produce many times more, which altogether changed the pattern of human life. This was linked with domestication of wild animals like cattle and horse and these Agricultural practices solved food problem of *Homo sapiens*, in most dramatic way, enabling man to produce and store food for crisis period. This led to development of human dwellings, near water sources, mainly rivers and development of more densely populated and stratified societies; Indeed, historians and anthropologists have long argued that development of agriculture made human civilization possible by concentrating some of them towards literature, architecture, arts, framing social laws etc, being free from food problem (7).

In understanding the universe , man started believing in an Almighty power, who has created the world. This Power or God was cursing the man by inflicting a disease on him, if something wrong was done by him. Thus, priests occupied a special status in the society who were responsible for conducting prayers etc for welfare of human population. Beside these priests, others also attempted to soothe human miseries. Centuries ago, Dhanvantari, Carak, Susruta (8), the well known Indian sages, developed Ayurveda or science of life as a subsidiary of Veda. Though Carak recognized germs responsible for disease, he gave little importance to these and considered diseases are caused by misbalance of body constituents like *vata, pitta* and *kapha,* their seriousness as well as curability to be pre-determined on the basis of the acts of past life; he was a firm believer in rebirth (8) and considered problems like leprosy as the curse due to sins done by man in his past life. Perhaps, this thinking deviated the scientific developments of East *vis a vis* those carried out by the western philosophers. We can also say that till that time man has not excluded past sins of man as main or one important reason for his ailments. Experiences of man with small pox started changing his thinking as following events revealed.

As per an estimate, small pox appeared around 10,000 BC and obviously it was considered a great curse of God for sins of man (9). Suffering of man by this ailment in India was termed '*devi nikali hai*" in general language showing importance of goddess and man's deeds for the suffering. Notwithstanding these facts, it became a common knowledge, as early as 430 BC, that survivors of small pox were protected from further exposures. With passage of time and long before 18th century man was practicing "inoculation" method in India, Africa and China to protect himself from small pox; in 'inoculation' method fresh matter from ripe pustule of smallpox sufferer was transferred to a susceptible man using a lancet (this was merely by scratching skin). This inoculation method came to Europe in the beginning of eighteenth century with the arrival of travelers from Istanbul and termed as Variolation as in Europe, smallpox was called "Variola" -derived from latin word "Varius" meaning stained or varus means mark on skin ; small pockes where pocke revealed sac and it was smaller in size (in comparison to syphilis). Though variolation was associated with 2-3% deaths, yet it was most powerful tool to protect humans from the dreaded smallpox and came in general practice in whole of Europe in early eighteenth century.

However, there was a general observation that dairymaids either did not suffer from small pox or mildly in comparison to other public. There was a general belief that dairy maids were in someway protected from small pox because of cow pox. And this was Edward Jenner ,born on 17th May, 1749 ,who took a lead to test this general belief or hypothesis. In May, 1796, Edward found cowpox lesions on the arms of his dairymaid, Sarah Nelms and he inoculated this matter of cowpox on 14th May, 1796 to a 8 year old boy James Phipps (10). In July, 1796, Edward inoculated the boy again but with small pox matter and no disease developed hence Edward concluded that boy developed complete protection against small pox. Interestingly, when Edward sent these findings in 1797 to the Royal Society for publication as short communication, the paper was rejected (a lesson for new aspirants, never to dishearten by rejection of papers from the editors). Jenner conducted some more cases and instead of sending his findings in the form of a research paper, he published them, in 1798, as a small booklet entitled "An Inquiry into the causes and effects of the variolae vaccinae, a disease discovered in some of the western countries of England, particularly Gloucestershire and known by the name of cow pox". Jenner used the word vaccinae as in latin vacca means cow and cowpox vaccinia ; therefore Jenner decided to call this new procedure as "Vaccination". As always happens with new discoveries, publication of the Inquiry was met with mixed reactions in the medical community. However, its importance was fast recognized by the medical community and general public ; British Parliament awarded British Pounds 10,000 in 1802 and £ 20,000 to Jenner five years later for his contributions. His works have been said to have saved more lives than the work of any other man and is responsible for development of a whole science of Immunology where he is regarded as 'Father of immunology'.

It must be emphasized that till the time of Edward Jenner experimentation, Germ theory of infections was not known (see chapter 6) hence there was no question of studying virus, associated with these infections. This germ theory of infection was firmly established by Louis Pasteur, a French scientist when he delivered his famous lecture " On the extension of the germ theory to the etiology of certain common diseases " on May 3, 1880 before French Academy of Sciences (11). Nevertheless, it may be concluded that Jenner's work was the first scientific attempt to control an infectious disease by deliberate use of vaccination. And this event was, perhaps, the start of thinking man that he can cure humans without taking much help from the God. Can we consider it a turning point in human thinking towards manipulating his own business with out the help of outside agencies?

The greatest change in concept of human life and on different creatures of the world was brought by British Biologist Charles Robert Darwin (1809-1882) through his revolutionary book "Origin of species" published in November 1859 (12). This book, caused most 'intolerant opposition' and produced great emotional controversy as it transformed whole concepts of human beings regarding nature and God and provided clear evidence that the plants and animals , existing at present, have not been created separately in their present form but must have evolved by slow transformation, natural selection and by survival of the fittest. It strongly advocated that this is the nature and natural laws which are responsible for origin of different species of plants and animals else life, on earth, started in a simple form. Even within the same species, there is intra specific selection in the form of sexual selection, where males of same species compete among themselves and most competitive get success in producing its progeny. The theory propagated by Darwin may be explained in simpler form by formation of varieties in a species, which changes into a sub-species and later in a full fleshed separate species. There are many forces leading to these changes which take millions of years for exhibition.

The studies made by Darwin come under Biological sciences that deal with studies of animals and plants- all living creatures present in the universe fall under these two groups. Again animals is not a homogenous mass and has been grouped broadly as vertebrates and non-vertebrates. The vertebrates have fishes, amphibians, reptiles, birds, mammals; non vertebrates contain a large variety of arthropods, annelids, worms and microscopic creatures like, protozoa, mites etc. These animals not only differ in their structure or morphology but also in their food habits though all depend on organic matter as their food. Some require air for respiration, others use oxygen dissolved in water while still others do not need oxygen at all. Likewise, it is not necessary that all the animals need blood circulation for their survival. How these animals maintain their life on the earth has been a matter of great study. The man or *Homo sapiens* is only one among millions of animal species but has occupied central stage for investigation resulting in development of many scientific branches. Each organ of man, like brain, liver, heart, kidney, blood, bones etc have given rise a full fleshed scientific subject where scientists are devoting their life to understand intrinsic value of each aspect. The well being of man and his domestic animals has given rise to development of health sciences which include Ayurved, Allopathy, Homeopathy, Veterinary medicine; within each, there are many sciences or subjects like anatomy, physiology, biochemistry, pathology, pharmacology, microbiology, parasitology, medicine, surgery gynecology. Each of these subjects is so vast that a scientist has to select and devote his whole life by selecting only a tiny topic within it to find out its complexities and solutions.

Biological science does not restrict itself to animals but also includes plants which are studied under botany, a separate discipline, which is also as vast as animal sciences. Here also, there are many subjects like agronomy, horticulture, plant physiology, biochemistry, plant genetics, entomology, soil sciences, water management, pathology etc where a scholar has to attain expertise in any of these fields and may select a topic for his research work. It is the Agriculture where new researches are continuing to develop new varieties of crops, for higher yields or producing crops even under adverse climate. Researches in biotechnology have opened new vistas where by genetic manipulations new varieties of crops like Bt cotton are developed; even work is going on enabling the plants to produce antibodies and alike biological molecules that may be helpful in human ailments (13).

By above examples, it become evident that scientific research on any topic becomes complicated, and it is not solved by one expert but a group of experts whose combined effort is essential to solve a scientific problem. This also reflects that for solving a biological problem, expertise of geographies, climatologist, geologist, mathematician, so distant from biological sciences, are also required. And it holds true for all biological subjects.

How research has influenced human life can be assessed by reviewing human civilization; its not more than five hundred years which have totally transformed human life. What so ever great changes are observed, are done during this period-starting from electricity, printing to rail, aero plane, computer, internet, mobile phones ; or a great many changes in biological sciences, starting from development of antibiotics against most treated diseases to rise of whole pharmaceutical industry, creating life by most unconventional method of cloning two cells, and understanding importance of stem cells in biological development as well as its potentials in treating many more human deformities. These innovations highlight the importance of research in the life of human beings and that no human race or a country can progress by remaining alien to research hence it is essential that our most talented, brilliant youth make research as their career.

3

Research and Your Personality

The very basics in one's life is whether he has judged his interests or decided goals of his life or has taken life casually. The first is to know what you wish to achieve in your life or in the words of our former president and Bharat Ratna APJ Abdul Kalam (14) what are your dreams ? If you have dreams than only you can achieve them hence it's essential to know what is your passion ?, passion is the key to success. In pursuing your passion, all things become secondary ; in fact, desired changes occur within you, unnoticed and automatic to achieve your goals. It may not be out of contest to mention that basically I belonged to a business family and being eldest son in the family, my father asked me to join family business, after doing my intermediate, with the liberty to join part time BA classes while carrying business activities. But business was not my passion hence I struggled to join to the academics and carried all my Veterinary education with distinctions - because of my interests. I am not sure what would have happened to my life have I joined least interested business line- perhaps a failed business man.

Nevertheless, there are certain characters in a person that facilitate research making a good choice for the career. If you possess these, they will strengthen you to enjoy research ; in their absence, your passion to research will metamorphose your personality to fit in research atmosphere. But in absence of both, it's better to realize that you are not fit for conducting research and there are all chances that you may repent in your later life. Herewith, we are discussing general characters of a scholar that help in research activities. However, this discussion is not about psychological tests but those which may play important role in your research career.

CURIOSITY

Admittedly, curiosity or to know the unknown is the nature of human beings and this and only this quality is responsible for all the developments that have taken place in whole of human civilization which we are witnessing today. Curiosity starts with the childhood when a child wishes to verify each and every item which comes in his way. But slowly, this curiosity is replaced by dumbness for not getting answers of his queries. Our education system is also responsible for causing fatality of this curiosity where a child is scolded for every question and rewarded to keep quiet and to give reply, in the examinations, on dotted lines for the questions he is asked. No doubt, our education system has miserably failed to inculcate curiosity in a student and has overemphasized gaining bookish knowledge with little scope for awarding new ideas searched by the students. This is the reason why our university toppers are those who imbibed bookish knowledge without its critical analysis. This is also the reason why most of these toppers have failed to shine in the research field which requires original critical thinking of a man.

Generally Indian culture is blamed for absence of curiosity in our students with the remarks that our culture does no permit to ask questions from our elders. However, if we scrutinize Indian heritage whether it has prohibited arguments, the outcome is just contradictory-and this has well be argued by Nobel laureate Amrtya Sen in his classical book "Argumentable Indian" (15). This is the Indian culture which condemned monarchy since long and promoted democracy where decision rested not on an individual but a group of ministers. Whenever something wrong occurred in the society,the Indian culture opposed these social evils with a result these were discarded by the majority. Again, from the time of Vedic era, it witnessed descent from time to time from among Indian populace and as a result there was spurting of different contrasting theories and religions- it may be Buddhism or Jainism or Sikhism. It was the Indians who ventured in different unknown geographies and created inventions which proved first for human civilization. Therefore, it is erroneous to blame that Indian culture prohibited arguments or suppressed curiosity in human beings. At most, we may consider it a gift of nineteenth century with multiple reasons that are responsible for killing curiosity in our society. We should not forget that the present education system was started by British Raj whose main aim of education was not to inculcate critical analysis or curiosity but to create clerks who can work well in English language. Sadly, we have miserably failed to change our education system as per our needs, even after independence.

This character to know almost all about the event or a matter or a living creature is the key of success to a researcher. The research starts with very popular words like what, how and when which our science teachers taught us in the schools. This is also linked with the character not to consider any written document as gospel truth or last word of science. If a scholar considers all the findings, written in our books, as true and mug up using his memory, he may prove a good student, securing highest marks in the examination, but will not prove a good research worker as this will not auger curiosity in him.

This curiosity should lead to know more about the subject. There are many questions which float in the air and scholars become interested to find their answer- and this is the way how science made advances. Why one plant survives for thousand years and others only a few days ? Why animals live shorter life than the plants ? Why mosquito completes its life as egg, larva, pupa, imago and adult while a louse gives rise to egg, nymph and adult. Why drinking semen by a female will never lead her to be pregnant or why man cannot change himself to lion and back to man ? How, organs are differentiated in an embryo which is the result of fusion of only two cells. Why large human populations suffer from a disease ? These and other questions led to advancement of science and the answers, based on scientific principles,were accepted by the scientific world. Yet there are many deeper questions which are puzzling the scientists and answers of some of them are opening new vista of sciences. Stem cell biology, creation of a sheep by cloning of two cells, development of genetically modified crops are some examples which have resulted by the new research.

Nevertheless, curiosity should not lead to arrogance which is quite different from the former. Neither curiosity should reflect to disrespect to any one - in fact it does not mean that whatsoever you are thinking is right and previous ideas were totally wrong ; rather you should think the reasons for this early wrong notion and this will help you to find truth which may be in the midway of the two. If you are really a true seeker of the truth, your behavior will be far from arrogance. Contrarily, you will start realizing journey of knowledge in human history and how little a man could know in his life span or how little man still know about the universe. When you are seeking answers from your seniors, there should not be a thread of arrogance in your behavior considering him poor in present knowledge or that you know better than him. When embarked with such ideas, it's better to think that what your senior is saying was true some years back, now your version may be true but there is no guarantee that it is an absolute truth -not to be challenged by any one in future.

CRITICAL APPROACH

As we have mentioned earlier, the outcome of all the creative activities is the result of individual's talents. For instance, the same canvass, brush and colors will produce different paintings, depending on the artist -how he is dealing the subject in his mind and reproducing the same on the canvass. This will also decide the value of the paintings. This is the reason why "Mona Lisa" or "The Last Supper", painted by Leonardo da Vinci in fifteenth century are still admired by art lovers. So is the case with scientific research. The outcome will depend on your knowledge of the subject,studying earlier works, what was left to be tested and how new approach is made to tackle the old problem. Taking all these facts in account, how you have designed your scientific experiment.

Obviously, if you are working on an old problem without adding any new technology or idea - in other words without critical analysis,there is very little chance to invent some thing new -whatsoever small it may be. And this is true even for small problems. If you have no new idea to deal an old problem, how the results will be different from the previous findings ?

When the old problem does not find its solution following traditional methods, scientists started fusing the two subjects to find new solution. They have achieved many new results by combining the knowledge of other scientific subjects with their own. The development of CAT scanning or sonography for human body is a beautiful example how knowledge of physics and biology were amalgamated together to find a new solution for an old problem.

We may cite here an example from Parasitology to narrate how a problem may be solved with little facilities but employing some new ideas. Here, life cycle of the flukes like schistosomes require fresh water snails namely *Indoplanorbis exustus* and *Lymnaea luteola* hence their maintenance in the laboratory is essential if we wish to work on any part of their life cycle. These snails were maintained, in earlier days, in earthen pots or glass jars or enamel tray by a tedious method of changing water daily which resulted in considerable mortality; moreover it required attending these snails daily including holidays (16). The rationale of this daily change was dependence of the snails on water dissolved oxygen which diminishes with passage of time; the snail feces also polluted the water hence change was needed.

With changing times, it became difficult to attend this daily routine hence we decided to develop a laboratory method of maintaining the snails where its daily attendance may be avoided without increasing its mortality. Deviating from early works, we started working on the factors which are affecting mortality of these snails in the laboratory. To our surprise, daily change of water was not good for the snails and it led to considerable mortality though change of water is required after a week or fortnight-depending on quantity of water used per snail- at least 50 ml of water per snail was required for changing the water at fortnight intervals. As lettuce (a good snail feed) was scare, other snail feeds were tried and ultimately mulberry leaves proved ideal for the snails, without producing sulpher di oxide which is lethal to the snails. The leaves are half boiled or steamed for young snails while adults consume any type of mulberry leave. The aquatic weeds increased dissolved aquatic oxygen, useful for snails, hence it's addition increased longevity of snails in the laboratory.

During this work in 1995 and after wards, the polythene bags have flooded Indian markets,therefore, we tried to keep the snails in batches in these polythene bags,filled with water at the rate of 50ml/snail, adding aquatic weeds and mulberry leaves. This method of keeping the snails in the laboratory proved simple, reduced mortality and occupied less space in keeping the snails in the laboratory with added advantage of separating them as per our requirements. Thus an old method of keeping the snails by changing their water daily was changed to a less cumbersome cost-effective method of keeping them without attending them daily (17).

Above example also suggested that if new techniques are not available in our laboratory, it's better not to initiate the work since we will fail to conclude our results. For example, our epidemiological works on nasal schistosomiasis have indicated epidemiological differences of the parasite in different geographies. For instance, at Jabalpur, nasal schistosomiasis or *Schistosoma nasale* is existing mainly in buffaloes without producing any symptoms, and in cross bred cattle, exhibiting clinical symptoms but local cattle are negative for the parasite when their nasal discharges were examined ; in contrast, nasal schistosomiasis is exhibiting clinical symptoms in local cattle at Balaghat district of Madhya Pradesh ; at Bhandara district of Maharashtra, *S.nasale* is causing symptoms in local cattle but goats are negative which is not the case with Tamil Nadu infection (2). All these epidemiological variations of different geographical areas are strongly suggesting possibilities of existence of different strains of *S.nasale* which are responsible for these varied results.

However, strain hypothesis can be verified only under controlled experimental plan. This will require collection of infecting material (positive snails) from all these places and infecting groups of genetically defined laboratory animals with counted number of schistosome cercariae and then recording all the biological details (cercarial penetration, percentage of development, ratio between mature and immature flukes, male and female, egg load in each organ, percentage of viable eggs, granuloma number and size etc) of each group to demonstrate existence of biological variations among different isolates. If the experiment is not done in genetically defined animals, its difficult to ascribe the variations due to the parasite. The experiment can be done alternatively by infecting genetically defined snails (*Indoplanorbis exustus*) with the miracidia, collected from fluke eggs from all these different places and then recording biological parameters like percentage of miracidia penetration, snail mortality at different time intervals, percentage of positive snails, incubation period, patent period, cercarial emergence per day and total cercarial output per snail etc Another supportive experiment will be applying PCR technique and to prove that these isolates differ genetically hence are of different strains of same parasite species- but PCR alone will fail to prove biological variations among the strains. In absence of any of these facilities, the results of the experiment will not be considered authentic and will be disputed in scientific world.

Therefore, at times, it may be futile to undertake experiments, if proper facilities are not available at the host institute.

KEEN OBSERVATION

Keen observation is another necessity for good research. A researcher should have the power to observe the event or any organism with total concentration and in different dimensions.You will note that this was keen observation which led to success of Archimedes or Newton or Alexander Fleming (antibiotic fame) otherwise the events might have lapsed as normal without giving any importance. The apple is falling from the tree since many years and will continue to fall but there was Newton who gave a serious thought to the phenomenon and provided an important law of gravity to the scientific world. This hints at one important fact that what so ever occurring in the nature, is occurring under certain law and only a keen observation may reveal mystery of the nature.

In undertaking experimental work, a prudent view is necessary while making observations, particularly if you wish to verify or nullify previous observations or hypothesis. For instance, Fairley and his contemporaries infected guinea pigs with cercariae of *Schistosoma spindale* and could recover only male flukes with absence of female flukes or their eggs hence propagated the hypothesis that the final host (guinea pig) has certain factors which did not permit selectively development of female schistosomes ; thus both intermediate host (snails) and final host (guinea pigs) have the capabilities to control sex of the schistosomes. This suggestion was having far reaching ramifications as it's the female and its eggs which are most pathogenic in final hosts and to find the factors controlling female development meant controlling pathogenecity as well as life cycle of the blood fluke. Therefore, the findings were widely appreciated with the need to furtherance this work.

About thirty years afterwards, Dutt (1957) designed an experiment (18) to verify if guinea pigs have certain factors that may check development of the female blood flukes and he took every precaution to eliminate the errors. Thus, the cercariae were collected from large number of positive snails (a single snail may shed unisexual cercariae), a large number of guinea pigs (30) were infected with varied number of the cercariae and the animals were sacrificed at different time intervals. During post-mortem of the animals, all the parasites were collected carefully from the animals and their tissues were examined microscopically for presence of *S.spindale* eggs. As the female blood fluke is slender with chances of its missing with naked eyes, all the blood flukes were examined for their sexes under a stereoscopic microscope ; and this microscopic examination confirmed presence of adult female blood flukes though with lesser number than the males. Further, the tissues revealed presence of the eggs though at times with absence of the females, suggesting its earlier death. Interestingly, there were some guinea pigs which developed only male flukes with absence of females or their eggs and if the experiment would have been conducted with lesser number of guinea pigs, there was a chance of encountering again only male blood flukes, thereby confirming a past observation- however false it would have been. This example is ample proof to suggest how a good planning and keen observation are important for an experiment to derive some fruitful results.

If you read keenly the discovery of Sir Ronald Ross (19) of incorporating mosquitoes in the life cycle of malaria (which received highest award in form of Nobel prize in 1902) you will notice that it was only the keen observation in last two experimentally infected mosquitoes where he could observe circular cells of about 12 micron size in the stomach of mosquito which happened to be oocysts of malaria parasite. And this keen observation was able to declare conclusively,first time, that insects are playing a role in transmitting parasitic infections. Prior that, he met a number of failures, also because of trying with *Culex* mosquitoes instead of *Anopheles*. Certainly, if keen observations have not been made in these experiments, there were all chances of missing an important finding and thereby revealing an important scientific truth. There is also the history of another Noble laureate Prof Alphonse Laveran (awarded Noble prize in 1907) who discovered malaria parasites in 1880es in human blood as black granules (later known as Laveran's bodies) even when blood stains were not in use (methylene blue was first tried in 1899). These were their keen observations and faith in their observations which led to such path breaking discoveries.

DEVOTION TO PROBLEM OR BEING CRAZY

This is an important character of all the scientists who have made great achievements in their life where scientific world still remember their contributions. You should have full devotion to the scientific problem you are dealing with. This character makes the scientist crazy and unworldly in the eyes of elite societies as many times the scientist, busy in his own ideas, forgets his surroundings and remains busy in his own thoughts. There are many stories about the great scientist Einstein (20) revealing his forgetfulness.

This character develops when a person is totally involved in his problem as his unconscious mind continues to find possible solution for the problem. And when all of a sudden he discovers the solution, he cries and run on the streets proclaiming " Eureka ! Eureka !" This is the reason why scientists have an image of being crazy or half mad in the society. They are so much devoted to their cause and continue to think about the problem even when performing their daily routines, that at times, they forget even social decorum. You may not find them well dressed in social gatherings or some other occasions. Indeed this holds true in all the professions which are related to creativity, may be literature, art, music or science. We have witnessed running senior doctors in night dress and sleepers to attend their emergency cases at Christian medical college, Vellore ;at that time, nothing is important than their clinical cases. You may visit and see the atmosphere of an artist who is busy with his canvass downloading his images or a literary giant busy in writing a book or an article. If you are always conscious about your dress or well being, you may not prove a devoted researcher rather you are an employee enjoying your life on the post of a scientist. Therefore, a true scientist should not care too much of outside world or its aristocracy.

This craziness is linked with the stamina you have to pursue your problem. If you are solving a big problem, you will never get its readymade answer; rather there will be many failures at initial stages; and if you do not have persistency and will power to pursue the problem even after failures, its very difficult to find the solution. Good research is always to solve a difficult problem which requires long attention. It may take a whole life to solve the problem and scientist has to remain committed totally to the problem. We are having examples where life efforts have failed to provide a practical solution to a chronic problem; one example may be of developing a vaccine against Malaria and trypanosomiasis.

This initial failure may be by not dealing with correct matter or agent. As the story goes on that Sir Ronald Ross could not get success initially for his theory of transmission of malaria by mosquitoes as he was using *Culex* species instead of *Anopheles* mosquitoes. Again, he could not get success,instantly, with *Anopheles* too, and it was only in last two mosquitoes where he could be able to observe malaria parasites. He confirmed his research by conducting his experiments with bird malaria. Therefore, this is only your craziness to solve the problem which ultimately provides victory.

Related to the above, is the problem of facing another type of failure when your experiments do not support your well cherished hypothesis hinting opposite facts.Such a situation may arise even in well equipped laboratory where there is no question of fewer facilities or funds but all your results are suggesting just opposite to your previous thoughts. If it is so, this is the correct time to analyze the whole experiments and to find the reasons for the difference- the basis of the previous experiments which led you to formulate the hypothesis and all details of the present experiments which are hinting otherwise. No doubt this situation gives cause of concern but demands diversion in different direction for re-checking the facts and circumstances where two types of the results are obtained. If your experimental plans were correct in both cases but the experiments were planned differently (e.g. using different rout of infection, animal species, period of observation, parameters selected), it may be that a third hypothesis, a combination of the two, may emerge with justifications for correctness of all the experiments, but with different models.

READING HABITS

If modern time has caused causality of anything, it is the reading and writing habits of our young generation–the two important requirements in research career. Even in higher classes, we are not witnessing reading of the text books by our students - rather they prefer to consult only class notes or guide books available in the market which do not deal the text but provide answers of some important questions. Though it was presumed that with introduction of objective question papers, a student will thoroughly study his text book and then only he will be able to answer the questions correctly; sadly, this has not happened and students are searching guide books with multiple choice questions and answers without bothering to read the full text.

Indeed, all our literary stalwarts and eminent scientists are voracious readers with habit of reading the books even beyond their field of specialization; reading has been the hobby of many big personalities who utilized their leisure time in reading the latest books and journals. In yester years, novels have been a good source of time pass and were also responsible to inculcate the habit of reading of voluminous books. However, with advent of TV, this has become the big source of our time pass as well as source of information discarding reading of the books. As watching TV is a visual phenomenon, it has curtailed reading habits so much so that persons have confined reading only to few pages or daily news papers or weekly magazines and are afraid of reading voluminous books.

But reading of journals and books is a must to carry over a successful research career and there is no escape from this. Every month, new books and research journals are published and it's necessary to read them to remain updated on the new knowledge pouring every movement in the scientific world. No doubt, there are certain changes taking place in publication world also with passage of time like publication of e-books - means that the scientists have the choice of selecting books in print form or in e-form that may be down loaded on his lap-top computer or iPod, i-phone so that he may read it at anytime and anywhere. Almost all the important scientific journals are also published on line which may be checked by the scientist any time; moreover, there are many paid websites (SCIENCEDIRECT, MEDLINE, CABI ETC) which provide facilities of abstracting scientific journals, annual reports, conferences, books etc and to give alert message on your mobile or e-mail whenever a research paper or book of your interest is published. This does not mean that hard copies of research journals or abstracting journals are not available or are not subscribed by the universities. You may select any one form as per your choice but you have to read either of them to imbibe the knowledge; and this is not once in a year but a continuous phenomenon; therefore it is important of inculcating reading as a habit.

we have witnessed some people, particularly in our business society, who presume that reading of books is only for giving examinations and once a person has finished examination or has taken the degree, there should be no relation between him and the book. This approach is detrimental to research activity as without reading the books and journals, it's difficult to acquire new knowledge or what is going on in the scientific world. To such persons, research is not the field of choice as reading is not his hobby or a passion.

CHARACTER

When we refer character, it means your integrity, your stand for a cause which is just and your spirit to fight fallacy; whether greed and fear will change your decision or you are able to withstand for just cause. At first, there appears no relevance of this to an outstanding research which brings fame and name to a scientist. But a deeper look to the subject confirms it's the character which matters a lot and at times you may find yourself on a cross road where you have to take a stand which may decide your future for a long time.

It's not necessary that such an occasion will arise in your research career. It may happen that you pass your time and research smoothly without facing any big problem but circumstances may put you at a stage where your decision is not based on your research findings but your character which will decide the ultimate outcome.

This character becomes a deciding factor particularly when you are conducting research which influences market forces and great money is involved. Here we may cite examples of the research which proved how tobacco is detrimental to human health- either consumed orally or through the smoke. This is not that only recent years research have shown its bad effects on human health; this was way back, perhaps more than fifty years when scientists, after conducting various trials, found that tobacco is harmful to human health. As true finding would have caused a heavy loss to the tobacco industry, all the efforts were made to shield this scientific truth. This was done in various ways; the important one was by influencing the scientists who claimed ill effects of tobacco on human health. And when they could not be maneuvered, the industry hired their own fake scientists, paid them heavily merely to counter claim the original findings and furnishing false data to show that tobacco is not harmful to human health- rather other factors are responsible for the ill effects, if any. However, these hired scientists could not alter the truth though were able to create a controversy; when more research data came from different sources over the ill effects of the tobacco on human health, it has to be accepted even by the industry which agreed to curve its consumption and of keeping away the products from the reach of children. There are many examples where scientific truth was suppressed either for money or religion but ultimately the truth prevailed and society hailed the scientists who withstood all sorts of pressures.

Perhaps, there is not such a great stack if you are pursuing research in Parasitology in India. This is also because discovery of drugs against parasitic diseases is not carried out in the country as most new drugs against parasites are discovered abroad ; therefore research on toxicity and efficacy of these products are also conducted outside the country except a recent trend where clinical research has been initiated in the country with new pharmaceutical products. However, there may be some ocassions if you are dealing with topics where health of large population is involved. For instance, Endosulfane, an insecticide used for paste control on agriculture crops, is found injurious to human health with reporting of many deformities including nervous disorders, hence its use in agriculture fields has been prohibited.

There is also use of insecticides for controlling ticks and other insects on our domestic animals ; as these insecticides are used externally, their fragrance may remain in the environment. It may happen that your research on any of these or a new insecticide conclude its harmful effects not only on the animals but also on the humans spraying the insecticide. Generally, harmful effects are visible only after a prolong use of these products hence it leads to a good escape to marketing companies unless a firm determination is made to prove otherwise and ask them to withdraw the product from the market.

There are Ayurveda companies who may be interested to market an herbal product against a parasitic infection and may approach you to conduct research for its efficacy against certain parasitic infections. It is good if your research finds the drug being efficient in killing the parasite but the problem arises when your research confirmed it's of no use against the parasite. The firm may request you to alter the result or at least not to publish the findings and the outcome is your character. A recent trend has been noticed that to market their products and to make them popular, certain firms have instituted some awards for the scientists who have worked on herbal drugs. And in many cases, the award and publicity goes to the work which highlights efficacy of these products.

An important dilemma comes when your research is contradicting the findings of an important scientist and verifications from different angles confirm the validity of your finding. Dealing such situation requires great patience and your integrity to the science. Generally, a true scientist will not take it otherwise, if you are contradicting his hypothesis ; but certainly your arguments should stand on sound fundamentals.It may not be a bad idea if you may discuss your findings with all the details to the peer scientist and to seek his opinion or criticism of your work so that you may have a chance to re-check your observations. But this does not mean to abstain you to publish your findings which withstand correct on scientific grounds.

OTHER FEATURES

We have discussed only some important features of your personality which are helpful in making research as your career. As this is not a book on psychology, we have avoided discussing other features though these may be helpful in your research career. An important character may be about your interests in social contacts. No doubt, the social contacts have no direct bearing with your research career. But if you are less social or somewhat more introwart, this may help you in reading or conducting your research experiments in the laboratory in isolation for hours together. This loneliness may be helpful in your serious thinking and conducting research without having concern for your friend circle.

Again, design of the experiment may require you to take observations continuously at certain time intervals and you may have to continue your work at odd hours including night hours. And you should be ready to take up the work with such a spirit. This is also while consulting the literature or writing a thesis or research paper as creativity cannot be restricted with time limits.

A common problem after getting sanction of a research project as principal investigator from a funding agency is its running successfully in an institute particularly if it lacks basic administrative infra structure. There will be problems in filling of project posts in time, getting sanction of imported chemicals where low quality indigenous chemicals are available at lower cost, or opening of L/C (letter of credit) for importing equipment and its clearance from custom department. There will be problems in attending mandates of the project and your library may be quite poor,not subscribing any worthy journal. All these problems are quite common for Indian scientists working in Indian organizations, coupled with zealous attitude of fellow personnel but you should be able to work under such difficulties without losing cool or inviting health problems.

4

Essentiality of Degrees

At times, we hear our young generation questioning the need for obtaining degrees for conducting any research work. To some extend they are right. If we look back, there are many examples where persons without any former degree attained what is impossible by many degree holders. Tulsidas has written *'Ramacharitmanas'*, incorporating so many principles of Vedas and stories from *Puranas* without obtaining any degree. Srinivasa Ramanujan (21) narrated mathematics theorems without any degree in mathematics. His achievements may be appreciated by the fact that we celebrate his birthday 22nd December as National Mathematical day. Our Nobel laureate Rabindranath Tagore (22) was without any university degree though in later part of his life, after bestowing Nobel prize, he was conferred honorary doctorate degrees from many universities. The best example in biological sciences may be cited of Nobel laureate Sir Ronald Ross, (23) who obtained a medical degree but was devoid of any research degree in natural sciences. In our contemporary period, one can cite the example of Steve Jobs (24) who,In 1976, along with Wozniak founded Apple Computer in the Jobs family garage and started marketing first personal computer making both founders millionaires in shortest period. Jobs also innovated appliances like I-pod, I-Mac, I-phone, which became darling of new generations due to multiple uses of the appliances. Mr. Steve Jobs, expired in Oct 2011, had not enjoyed a comfortable childhood and was a drop out of Reed College, Portland. The other famous example is of billionaire William Henry Bill Gates III (25) (born on 28th October, 1955) who remained number one in Forbs world rich list for many years due to his innovations in computer software and establishing famous Microsoft company in USA who launched various Microsoft operating platforms like Window 2000, 2007 and window 2008. Again he is a Harvard University dropout, confirming that degrees are not essential for achieving miraculous goals.

Though without any former degree, one thing, common in all these great personalities who changed the world, was their passion to their work and higher power of observations, which have culminated into the great works. Moreover, their memory was unquestionable; prior writing the great epic *Ramacharitmanas*, Tulsidas used to deliver '*Ram Katha*' incorporating so many episodes/couplets from *Veda* which he has heard from his Guru Shri Nari Harijee and Shesh Sanatanjee in his childhood or young age. How many scholars do remember their lessons or even important hypotheses with intrinsic problems after finishing their examinations in that particular course ?

As stated above, a university degree is not essential for conducting any type of research as research is the result of curiosity- to know the unknown; in fact, human civilization has progressed only with new innovations, many of which have been carried out by those who did not have any formal education. There are many examples in our day to day life where our so called laymen have developed a '*juggad*' tempo from assorted items to transport rural population or a bicycle which can generate electricity; our ICAR is awarding the farmers, who do not possess any former qualification, every year for new innovations, be it in fruit varieties, organic insecticide or in agricultural implements.

However, in the present era, it is the education which exposes the students to new world of advances in science or other disciplines. In that sense, you are lucky to have present education system which provides ample opportunities to assess knowledge from any source. This present education system is not more than 400 years old. Prior that, it was difficult to assess the books,as printing was not in operation, and acquiring knowledge was not so easy. Moreover, education was confined to rich persons or was caste based,restricting its assess to high class society. It will be interesting to know the plight of those who have no assess to important books in olden time.

In contrast, present education system has provided best opportunity to anyone to learn from the experiences of others in the form of consulting books written by these well known scholars. An additional advantage of the education is to confirm if your previous worker has already completed the innovation which you are intending to undertake.

Moreover, the modern science has become so complicated with so many hypotheses and use of advance equipments that it's very difficult to understand basics of these by any one who lacks a basic training in science. Therefore, it is rather impossible to make any important scientific achievement without taking training in that particular discipline.

You may conduct any type of research without any degree as per your instinct and ego but if you are interested to conduct research in any recognized institution, either in India or abroad, former qualifications are essential. You are to be judged by the peers whether you possess the required qualification or not for solving a research problem prior allowing you to join the organization. Perhaps, this is for two reasons; a degree confirms your credibility and capability of conducting research. Secondly the organization is paying you every month a salary without knowing the outcome. It is also incurring continuous expenses on the equipments and chemicals hence a justification is required in appointing any person whether he is capable to deliver the results. Therefore, obtaining a degree is essential prior joining any research organization or any university. Indeed, you cannot join any research organization, in present circumstances,without possessing the required qualification, experiences and competing in an interview.

These qualifications are needed in the related discipline, thus a doctorate in physics will be considered unsuitable in zoological research and vice versa. This is because of existence of vast knowledge in the subject which is impossible to imbibe without studying the particular discipline. Nevertheless, there is need of synthesizing the two distinct disciplines into one to achieve out standing research; thus new branches like bio-engineering or bio-physics have evolved and a new demand of scientists has been created in such unusual disciplines where such requirements are preferred.

If we analyze our education system, eventually the course syllabus, our text and reference books are bunch of testimony of experiences of our ancestors or contemporary scientists. As life is short to experiment all which previous personalities have experienced, our text books help us to acquire knowledge from their experiences. This also implies that what ever is written in the text book is not absolute truth and subject to change as per further experimentations. This holds true especially in all Science subjects. Our great Philosopher Swami Vivekananda (26) has echoed the same voice,while preaching, never to follow any one blindly and testify yourself prior accepting a truth.This perception is important for conducting research since many students do not try to think beyond what is written in the text books.

In our education system, High school teaching is to make students aware about languages, basic principles of sociology, mathematics and science. Thereafter, the student may choose Art, Commerce or Science streams. The science is bifurcated in Biological sciences (Biology, Physics, Chemistry) and Physical science (Physics, Chemistry, Mathematics). Now, mathematics has gained its importance in biological sciences as well, hence some schools, particularly central schools, are allowing biology students to take mathematics as an additional subject. The students can not change their disciplines at graduate level in science subjects due to specificity of the subjects though there is provision of exchange of subjects in other disciplines.

Since modern science has evolved on the scientific experimentations, practical is an essential part of all the science degrees. Indeed, without conducting practical, a student cannot understand the subject in its right perspective. Thus, if you are studying anatomy of a vertebrate or non-vertebrate, it is important to dissect yourself the body of the animal and trace out the systems- this is how our predecessors acquired knowledge of human anatomy and also related physiology.

At degree levels, there is the need of biological materials with some sophisticated equipments. If you want to acquire practical knowledge at this level, rest assure that the degree colleges are having good laboratory facilities else your scientific knowledge will not be accurate. There are some degree colleges where even basic facilities are not existing and students are unable to conduct basic experiments. In some degree colleges, there are only few slides to show and two,three common experiments which are either demonstrated or conducted by the students. If you have acquired a science degree from such a college, your basics will remain weak, hence will have to work hard for doing some fruitful research.

The case is more complicated at Postgraduate level, which means both Master's and Doctorate degrees. Here degrees may be obtained with little efforts and without acquiring practical knowledge or very hard work may be needed for clearing the degrees. If our aim is solely to obtain a Master's degree, the students pay attention only on those topics and conduct practical which are important for examination point of view; but those who dream to advance their career in research try to understand the whole subject with all possible ways. Generally reputed colleges, institutes, universities have good facilities, including latest equipments and an expanded library and desire hard work from their students,prior conferring these degrees. All graduate colleges are supposed to have their own museum which depicts collection of biological materials. This collection consists of specimens of different animal classes, depicting their characteristics, their dissected bodies for exposing internal structures, life cycles of birds, frogs to acquaint a student about basics of biological sciences. In Medical and Veterinary colleges, each department is maintaining its own collections, exhibiting important features of the subject; the more reputed college will show more rich museum.This is the reason, why more emphasis is given to the institute,from where a degree is obtained, during interview for research posts.

Doctorate degree is the highest degree in our education system and considered limit of the knowledge, a scholar is supposed to acquire. But from research point of view, this is the basic degree for research where a scholar learns how one can solve a problem or can provide his opinion or hypothesis on a given problem. Thus, this is not the end of the research but beginning of a research career and the scholar is supposed to explore the field of specialization with the hope to provide new solutions to existing problems.

The Agricultural Universities adopted American education system and made it compulsory to undergo course work also for acquiring a doctorate degree in Agriculture, Veterinary and Agriculture Engineering sciences. This is beside the thesis work they have to do on some problem. But such course work has not been essential in traditional universities. While guiding the students of traditional universities for their doctorate program, it is our experience that the scholars forget most of their subjects and remember only integrities of their thesis problems. Thus many of these Ph.D. holders were unable to clear UGC lecturer ship entrance test. In one sense, these Ph.D. degrees were made so cheap that any scholar may obtain it with little effort after finding a suitable guide. Perhaps, this lowering of the standards drew attention of the authorities and now University Grants Commission has made it mandatory to conduct an entrance test prior permitting a student for Ph.D. program; simultaneously, UGC has also incorporated course work in Ph.D. program so that students should remain aware about the latest advancements going on in their subjects (a part of the course is about guidelines for conducting research -as envisaged in this book). For a research career, it is essential to continue studying the subject throughout life and never abandon the idea of acquiring knowledge from what so ever source it is available.

DEGREES IN PARASITOLOGY

For joining a research institute or a university as faculty member in Parasitology, a master's degree in Parasitology is essential whereas a doctorate is preferred by most organizations. It's better to equip yourself with a Ph.D. prior joining your service career as this is the basic qualification for your promotions to higher posts. There are three main branches of Biological sciences i.e. Veterinary, Medical, Zoology where parasitological research is conducted and teaching to the students is imparted. Though all these three faculties are working on parasitological research, their field of work differs to some extent. Veterinary Parasitology deals with parasites and diseases caused in domestic, wild animals, domestic birds, and zoonotic parasites - which are transmitted both to man and animals. Medical Parasitology confines itself with the parasites of humans and their diseases including zoonotic parasites where as Parasitology as a part of zoology/life sciences deals any branch but mainly basic aspects of the parasites like taxonomy and life cycle. Perhaps, due to shortage of openings, there are restrictions for joining the post in respective faculty. Therefore, to join Parasitology department in a Veterinary institute, you need a master in Veterinary Parasitology where as medical institutes do not have separate departments of Parasitology and it falls either in department of Pathology or Microbiology. In research institutes of CSIR (e.g.CDRI, Lucknow), ICMR, Parasitologists of all faculties may apply and join the post through interview.

You may select the faculty as per your interests and basic qualifications. Its best to acquire degrees from a reputed institute/ university which is having eminent scientists/teachers in its faculty with modern research facilities as when you are engaging your two-three years or more for gaining experience and training in a particular discipline, let it be in a best institute with best environment. Please check credentials of the faculty of the department- both past and present; how many national and international awards have been conferred to them, how many are awarded fellowships of science academies and research projects running in the department; whether there is any collaboration with any international research institute- all these facts suggest about the position of the department and its standing in scientific world. We are herewith mentioning some good institutes which are imparting degrees or knowledge in Parasitology.

DEGREE IN PURE SCIENCE PARASITOLOGY

At graduate or BSc level, a student who selects Biology, is taught subjects like zoology, botany and chemistry where a general knowledge of the subject is imparted. It is only at post-graduate level where more specialization is done. The universities are running two years course in MSc zoology or life sciences and three years program of Ph. D- where the degree is awarded either in zoology or life sciences as the case may be. In India, Basic Parasitology is a part of MSc zoology or life sciences and taught in the universities as part of invertebrates whiles some universities arrange a special paper on Parasitology as a curriculum of MSc in zoology. Please remember that the students are not awarded a degree in Parasitology but only in Zoology or life sciences. The MSc program is solely depending on course work though some universities are asking to submit a dissertation on certain topic. Ph.D. was earlier solely depending on a research program leading to submission of a thesis to the university. Recently University Grants Commission has made it mandatory to include one year or six month advance courses to all those who are getting registration for Ph.D. program. Again an entrance test is conducted by the university for judging eligibility of the student for registration in doctorate program.

Earlier, Zoology faculty was dominating in research mainly as it was related to taxonomy,check list, and life cycle of the parasites. As most of the work has been done on these topics by traditional methods, the new research is shifted towards biochemistry of parasites, electron microscopy, evolution, behavior of parasites,environmental studies, molecular diagnosis, phylogeny or genetic aspects of parasites, zoo geology or geo biology; many new topics of life sciences are being worked out like neurobiology, aging, tissue repair mechanism, trigger mechanisms etc using invertebrates or parasites as models. However, there are only limited universities which are working on these new topics while others have stopped working on parasitic problems; therefore, it's better to check activities of the department prior taking admission if you are interested only in parasitic problems.

Though Indian traditional universities are not awarding degrees in Parasitology, a research scholar may select his research problem on the parasite provided his guide is also working on such topics. Clearly, basic research may not provide an immediate solution of an existing problem but its results are far reaching, affecting all aspects of parasitology. Even if a scholar is trained in other fields of life sciences, his interests may shift at later stage using parasites as their models. Hopefully the students may be exposed to many fascinating fields of life sciences in these universities which may ignite interest to take up this field for his future research program. We are herewith mentioning some Indian universities who have earlier contributed richly to our knowledge on the parasites.

Allahabad University : The Zoology department of Allahabad University was established in the year 1906 and is one of the oldest departments in the country. The department has a very good zoological museum with some rare invertebrate specimens. Dr AD Imms and Dr SC Verma have worked extensively in Entomology; Dr HR Mehra, an important figure in Parasitology research, introduced Helminthology in the department in 1925 and produced many eminent students who later headed helminthological research in the country. The university has received its central status paving the way of higher research facilities. The university has a separate Parasitology laboratory and offering Parasitology as a special subject to the MSc students in their final year.

Aligarh Muslim University : The faculty has worked extensively and gained international fame by working on nematodes, particularly filariids with distinguished scientists like Dr Babbar Mirza and Jairajpuri who later joined as Director of Zoological Survey of India,Kolkata.

Andhra University : It has two campuses- one at Warangal and other at Vishakhapatnam. The faculty of zoology has worked extensively on the parasites,especially on histochemistry of helminthes and provided lead in this field. Drs Hanumantha Rao and Madhavi have been distinguished Helminthologists of the faculty.

Lucknow University : The Zoology department came in existence in 1921. The department has been famous for classical zoology research like helminth taxonomy,entomology and toxicology. The faculty had eminent scientists like Drs GS Thapar, Premwati, SC Baugh who have pioneered on parasites of domestic animals and birds. The department has a good central instrumentation laboratory and well equipped computer laboratory.

FOREIGN DEGREES IN PARASITOLOGY

As stated earlier, none of these or any other traditional Indian university is providing MSc degree in Parasitology, although India,being a tropical country, is facing myriad parasitic problems and there is a need of a national institute on Parasitology (27) to train young scientists in the subject. Strange it may be but its true that some European and American Universities, where parasite is not a big problem because of cold climate, are providing degrees in Parasitology where a biology student is also eligible to be admitted. Here are some examples:

The University of Copenhagen, Denmark is offering a two year degree course in MSc Parasitology. The admissions are made in January and October and open for zoology, veterinary and medical students with a bachelor's degree in the subject.The course is taught in English and thesis is required prior awarding the degree.

There are at-least 7 universities in UK which are imparting Master's degree in Parasitology. The course duration varies from one to two years where submission of a dissertation is required. The Liverpool school of Hygiene and tropical medicine, UK was joined by Sir Ronald Ross when he settled in Britain after conducting pioneer research on Malaria at Hyderabad, India. Later, he joined London school of tropical medicine and Hygiene,till his death in 1932. It may be worth to mention that the latter institute was established by Prof Patrick Manson who is considered father of tropical diseases and was a mentor of Ross. Both the institutes are imparting Master and doctorate degrees in Parasitology with scope for advance research on parasitic diseases. Some other important universities are :

a) University of Liverpool

b) University of Sanford, Manchester

c) University of Nottingham

d) University of Manchester,UK.

There are 6 institutes in Malaysia and 8 Universities in USA which are imparting teaching and research in Parasitology. A few universities in USA are :

i) Iowa State University

ii) University of Florida

iii) Texas A & M university.

All the above universities are providing courses in Parasitology where zoology or life science scholars are also eligible. Beside there are Agriculture universities with Veterinary Colleges in these countries which are imparting courses in Veterinary parasitology. The details may be downloaded from Google or any other website. The international students, including Indians have to clear English proficiency tests like TOEFL, IELTS, and PTE for taking admissions in these institutes. There are a number of scholarships, assistantships which students may compete for sustaining their livelihood during study period like Commonwealth scholarships, ICAR international fellowships, Rhodes scholarships and many others.

VETERINARY PARASITOLOGY

The Veterinary colleges were started in India to look after animal health and related subjects. Prior to independence and thereafter, these colleges were imparting teaching up to diploma or graduate level (meanwhile, some colleges initiated post graduate programs) and were under administrative control of state animal husbandry departments. When Agriculture Universities were established in 1960 and onwards in the country, these veterinary colleges became constituent colleges of these Agriculture Universities which followed American land grant pattern with semester system of the education. Formation of Agriculture Universities gave an impetus to research with start of master degree program in veterinary subjects with first attention paid to Pathology (including bacteriology and Parasitology as part of its curriculum), Nutrition and Genetics. Soon Ph.D. programs were also started in these subjects; Madras Veterinary college, Bombay Veterinary college, Mathura Veterinary college and Jabalpur Veterinary college were the first who started masters'/doctorate programs in these subjects in the beginning of 1960es. As science advanced, there was formation of first Veterinary University in Tamil Nadu state in the year 1989 and in the present times almost all the states are having Veterinary university divesting the Agriculture university from the faculty of Veterinary and animal sciences. It is heartening that all veterinary colleges have Parasitology as a separate department that is imparting MVSc in Parasitology as well as many colleges are running Ph.D. in Parasitology. In India, only Veterinary colleges are having separate departments of Parasitology and are imparting post graduate degrees in Parasitology otherwise neither zoology/life sciences nor medical sciences colleges have separate department of parasitology or awarding degrees in Parasitology.

MVSc in Parasitology is a two years degree program where BVSc is essential for admission. It is not necessary that past distinctions of the departments are still maintained. This is because of retirement of senior faculty members resulting in shortage of staff in many colleges; even some colleges are running their department and courses with presence of one or two faculty members. Therefore, while seeking admission, it is imperative to enquire about the present position of staff, its library as well as about equipments and research projects being run by the department. Here, we are mentioning name of some universities where appropriate facilities are existing but in their old Veterinary colleges while newly established colleges are still struggling for staff and modern equipments. The id of their website is also provided so that details of veterinary colleges may be searched by the prosperous students.

1. Indian Veterinary Research Institute, Izatnagar, Uttar Pradesh. *www.ivri.nic.in*
2. Govind Ballabh Pant university of Agriculture and Technology, Pantnagar, Uttarakhand *www.gbpuat.ac.in*
3. Tamil Nadu Veterinary and Animal Science University, Chennai *www.tanuvas.org.in*
4. Pandit Deen Dayal Upadhaya Pashu Chikitsa vigyan Vishwa Vidyalaya evam Gou Anusanthan Sansthan, Mathura, Uttar Pradesh *www.upvetuniv.edu.in*
5. Nanajee Deshmukh Veterinary University, Jabalpur, Madhya Pradesh. *www.mppcvv.org*

6. West Bengal University of Animal and Fishery sciences, Kolkatta, West Bengal. *www.wbuafsci.ac.in*
7. Lala Lajpat Rai University of Veterinary and Animal Sciences, Hisar. *www.luvas.edu.in*
8. Karnataka Veterinary Animal and Fisheries Sciences University, Bidar, Karnataka *www.kvafsu.kar.nic.in*
9. Orissa University of Agriculture and Technology, Bhubaneshwar *www.ouat.ac.in*
10. Guru Angad Dev Veterinary and Animal Science University, Ludhiana. *www.gadvasu.in*

MEDICAL PARASITOLOGY

Of late, there is opening of Medical University in Tamil Nadu in 1988 which has triggered start of Medical Universities in other states too, though many states still lack a Medical University which can monitor medical education. These Medical Universities are providing administrative control of state Medical colleges and are supposed to encourage high class research in medical sciences. India, being a tropical country, is facing great health problems due to parasitic diseases, but its unfortunate that state medical colleges do not have separate department of Parasitology and it is still taught as a part of Microbiology which is again a part of Pathology; the post graduate degree i.e. MD is not awarded in Parasitology but in Microbiology or Pathology though a student may take up his research problem on a parasitic disease. Still there is no inclination of medical graduates to join post graduation in Microbiology or non-clinical subjects resulting in shortage of teaching or research staff in these subjects.

Visualizing these problems, some central medical institutes i.e. AIIMS, New Delhi, JIPMER, Pondicherry, PGI, Chandigarh have started post graduations in non-clinical subjects where a basic medical qualification is not essential for getting admission. These institutes are admitting non-medical students in the subjects like Physiology, Biochemistry, Anatomy etc but there is no medical institute in India that might be imparting MSc in Parasitology. How contrasting is the situation of this tropical country, where parasitic diseases are in abundance, *vis a vis* American or European countries which are running separate courses in Medical Parasitology but without facing such problems. This compels us to mention what Sir Ronald Ross has stated in 1931 in the preface to the 3rd edition of his book " In Exile" (28) "*But there seems to be an unfortunate apathy regarding great medical tragedies, namely that they are often allowed to pass by with little or no attention".*

It is only Post Graduate Institute of Medical Education and Research in Chandigarh which is having a separate department of Parasitology. However, the department is not imparting MSc Parasitology though doctorate program is run by the department where scholars from all three faculties are admitted.

Though there are no separate departments of Parasitology in medical institutes in India, the faculty of many of these institutes have realized importance of parasitic diseases and have started conducting research on these topics. Here, I will like to mention about All India Institute of Hygiene and Public Health, Kolkatta (www.alihph.gov.in), established in 1932, and happened to be the oldest school of public health in whole of South East Asia. This institute was started with the assistance of Rockefeller Foundation and have 11 academic departments with two field practice units; departments of epidemiology, microbiology, public health administration have some concern with parasitic diseases.The institute does not run degree in Parasitology but it runs Master of Veterinary Public Health (MVPH) where problems related to parasitic diseases are taken; eligibility for this course is MBBS or BVSc & AH degree holders.

To augment research work and for better collaboration, The Indian Academy of Tropical Parasitology has been established in JIPMER, Pondicherry where more and more medical microbiologists and other scientists, with interests in parasitic diseases, are becoming members of the academy. While delivering Dr SC Parija oration award lecture at Shri Auribindo Medical institute, Indore, I have advocated to start a National Institute of Parasitology (27) (see www.indianschistosomiasis.blogspot.com); this institute will impart MSc and PhD degrees in Parasitology to the scholars of all three faculties and will have mandate to tackle parasitic diseases of the country. Hopefully, there are chances that some changes will occur with passage of time when talented scholars will join this stream and will help in solving the parasitic problems of the country.

5

Job Opportunities

Whenever a person acquires desired qualifications, he needs a forum to express his talents and a place where he can fulfill his dreams and make important attainments. This may be achieved by two ways. One is, he becomes an entrepreneur and plans his own organization. As we are discussing biological research, there are good opportunities to start a Pharmaceutical firm to produce a drug, or a biotechnology firm to produce some biochemical molecule, enzymes, bio-fertilizers, vaccines against important human/animal/poultry diseases or a firm to supply diagnostic kits or immunological products etc. These are not hypothetical suggestions, difficult to achieve by a middle class person, but part of reality achieved by middle class family in India with laurels in their field of work. We may cite examples of firms like Dr Reddy Laboratory (Dr K.A Reddy), Biocon India (Ms Majumdar), Venky (poultry vaccines etc), Biorad (Bangalore based biochemical molecules supplier) who are working in India and leading their respective field. All this have been possible with tremendous change that has occurred in Indian scenario where there are many venture capitalists (or PE firms) who are willing to invest any amount of money provided they like your new idea of establishing a new enterprise. A somewhat more detailed information, in this regards, is provided elsewhere (see Research for money-chapter 13). Here I have not mentioned to pursue a Dairy or Poultry farm as these are not parts of research activities but related to animal production and any other person may also enter in such commercial activities.

Nevertheless, it is true that most of our educated middle class or Ph.D degree holders feel lack of courage or initiatives to initiate their own venture and are interested to get a job which can fulfill their desire and dreams. However, for seeking a suitable job, it is equally important to acquire a suitable qualification that makes you eligible for the job; interestingly, this is not essential for spreading your own venture but a must if you are seeking a suitable job which is directly related to your qualifications and experience.

Many times, a student seeks admission in some unusual courses with the hope of getting higher job opportunities which,at times, back fire. I may cite a few examples that reveal the problems one may face in acquiring traditional jobs with these unusual degrees. Biology students earlier had option either to select zoology or botany for their M.Sc degree. Later, many universities began offering M.Sc in Life Sciences and discontinued zoology or botany degrees. However, when teaching posts of the colleges were advertised, these were for zoology or botany with postgraduate qualifications in respective subjects, making life science degree holders as non-eligible.

In enthusiasm to start job-oriented courses, many Science colleges started graduate or post graduate courses in Pathology, medical laboratory technology, microbiology, Physiology etc but as these were not recognized by Indian Medical Council or other medical authorities, the students were neither eligible to open their pathology laboratory etc nor to apply for traditional teaching posts.

It is not a rosy picture in India with regards to job opportunities for students belonging to science discipline, particularly persons with basic science degrees. Of course, they are eligible to apply for any post where a degree or post graduate degree is required but there appears only limited jobs in their respective scientific fields,despite miniscule production of Ph.D. holders in India.

The biggest employer of science graduates is our education department- may be higher secondary school or colleges or universities. Today, India has one of the largest higher education system of the world with 431 universities (including 14 open universities), more than 21000 colleges, 14 million students and more than 505,000 academics as teachers (29). These institutes have failed to absorb even half of our talented scientific population; financial problems coming in recruiting faculty coupled with delay or no sanction for filling of vacant posts. Several years have passed when state Public Service Commission of many states have advertised teaching posts of degree colleges or universities thereby throwing these talented scholars in depression and everlasting unemployment period. As per a report (30) America is producing 25,000 Ph.D. holders per year, 35,000 are produced by China where as only 5,000 Ph.D. holders per year are produced by India. Therefore, it should be a matter of great concern that India is not able to provide suitable job even to this meager number of Ph.D. holders. It is therefore not surprising why a large number of science graduates change their field in the middle and try to be absorbed elsewhere or our Ph.D. scholars abandon research program in the middle to join a lower class post. It is a sad commentary that in present circumstances there is no incentive which can attract our talented students to join research as their service career or even assurance of getting a research job after completing their Ph.D.

As education department creates largest number of vacancies, most of our science post graduates join colleges in the beginning of their career.It may be clarified that these teaching posts of higher education are linked with research- at least as per rules and regulations of higher education. So much so, that Ph.D. has made essential for seeking a lecturer post in these institutes or you have to acquire the same within stipulated period. It is also made mandatory to publish research papers in peer reviewed journals for being eligible for higher posts. Thus it is important for every teacher to conduct some type of research to remain eligible for higher posts. There is no separate posts of teachers and researchers in post graduate colleges or traditional universities and it is the compulsion of the teachers to publish research papers for their survival. This importance to research is based on the assumption that imparting higher education requires some efforts to know the unknown by the teachers.

However, some professional colleges and Agriculture universities have two categories of the posts-teachers and researchers. They advertise certain research posts under some research schemes where the scientist has to devote his whole time to solve the specific problem though teachers are not excluded from the responsibility of conducting fruitful research. It is obvious that the persons do not apply to specific teaching or research post as per their interest but totally depending on availability of vacancy at that time.

If we discuss job opportunities related to Parasitology, this is a more specialized discipline and concerned with parasitic diseases of animals and human beings. Therefore, it may be presumed to have ample posts in public health and animal husbandry departments where a specialized degree in Parasitology is required. Ironically, this is not so and it may take some time to realize the need of Parasitologists in dealing with parasitic diseases.

In Zoology department, Parasitology is taught under invertebrates and all posts asking specializations in zoology or invertebrates are eligible for the persons holding such specialization. In Medical faculty, Parasitology is not a separate entity and comes under either Department of Microbiology or Department of Pathology. There are departments like Pathology, and Preventive Medicine which also considers parasitology as eligible degree for certain posts. Therefore, you are eligible for posts of these departments what so ever specialized work you have done. It is in the Veterinary colleges which have separate department of Parasitology where persons holding degrees only in Parasitology are eligible.

Animal Husbandry department do not have specialized posts requiring specializations,rather, higher posts (assistant director,deputy director, joint director) are awarded on promotion basis where seniority is main criterion. This is some what different in Public Health department which advertises higher posts requiring qualifications in Medical microbiology or Pathology or community medicine or other discipline.

Ironically, all the three faculties have created water tight compartments where the posts, in their respective faculty, require a basic qualification of their discipline; thus a person should hold BVSc and MVsc in Veterinary Parasitology to get a job in Veterinary faculty. Likewise, in Medical faculty a medical qualification is essential for most of the posts although there may be some posts where medical qualification may be relaxed.

There are research organizations like ICMR, CSIR, DBT, ICAR, DRDO, which advertise technical posts with essential research experience in specific research field of Parasitology, irrespective of their basic qualification.

There is a ray of hope with new changes in Companies Law,2014 where 5% net profit is to be expanded in CSR (companies social responsibilities) by the big firms. Hopefully, these firms will spend more money on education, environment and health through non-government organizations and other bodies. We hope this will open new job opportunities where Parasitologists will also be able to show their working strength.

MIGRATION TO OTHER COUNTRY

To above job opportunities, we may add opportunities talented scholars are having in European and other countries. Its true, that science is free from political boundaries and any person willing to pursue his scientific goal is free to migrate to any country where his dreams may be fulfilled. There are many examples in the history where scientists have migrated to other countries due to one or other reason and have shined by publishing new discoveries. The foremost name,among them, is of Albert Einstein, a German and noble laureate, who migrated to USA and worked on relativity and allegedly helped in Manhattan mission for developing atom bomb. The history is full of Indian talents who faced problems in India and migrated to other countries where their scientific work earned not only scientific acclaim but also noble prizes. We may name here Drs Har Govind Khurana, Ramakrishnan and alike.

In present day scenario, it may not be political compulsions but unemployment or under-employment, or less laboratory facilities or suppressing atmosphere that may compel you to migrate to other country for achieving your scientific pursuit. This migration is done at two stages-first where you have acquired Ph.D from India (may be some research experience also) and migrate after getting a job card from abroad. It may be easy to seek assistantship in a research project, if you are holding a Post graduate or doctorate qualification and have published one or two research papers which reflect your talents in dealing a research problem. There is no harm to migrate to any country at any stage if you are unable to get financial grant for your new research idea and if a scientist of other country is willing to support you in your scientific endeavor.

The other stage is for getting a degree from a foreign country and if conditions are suitable to seek a suitable job and settle there. For getting admission in other country, you must acquire at least a graduate degree from a good university of India as its very difficult to enter in another country without good academic records. Depending on your performance in academics, you may be allowed to join any ongoing research project of the department and rest of the story depends on your intelligence, diligence and your devotion towards science.

If we analyze the present scenario, it is a fact that western countries are having good scientific facilities, necessary for carrying experimental research. Their laboratories are possessing modern equipments with no shortage of costly chemicals and biological molecules needed for running these equipments. Their libraries are really a treatise of knowledge containing large collections of latest editions of the books and almost every journal of your field.

Perhaps, the most glaring difference between India and these countries is the mind setup in dealing scientific problems. While it may take more than a year to acquire a copy of a new book in India, the book will be on your table within a week without going through many rubbish formalities. And the most disturbing moment may come when your new research idea may fail to garner the finances from the Government agencies (ironically, almost all the finances are coming from Indian and state governments which have their own norms for sanctioning finances; despite all talks, private companies are lacking such generosity) as review committee has not appreciated your plan. These sanctioning bodies fail to appreciate that there may be some very new ideas, nascent to the subject, which deserve attention and trial by our young talented scientific pool.

No doubt the conditions are more favorable in western countries but there are some basic differences and it's better that you realize them sooner than later. The most important is the difference in the culture of the two societies and you have to prepare your mind to adjust in that culture amicably along with your family.Secondly, the western universities desire a scientist or a teacher to work till his retirement as there is no bonus for past laurels; you have to be on your toes always and should remain active throughout your academic life to survive in the atmosphere - just opposite to India.

Moreover, if you are migrating to an European or American country for doing research in Parasitology, it should be clear that there are many parasites and parasitic diseases which are not prevalent in those countries hence there are difficulties in taking up epidemiological and related work on these parasites, though some universities have international projects with the provision of visiting and working in endemic countries. Therefore, most of their work is related to molecular Parasitology or basic problems on the subject. If your aim is to work on epidemiology or diagnosis etc on parasitic diseases, it's better to stick to a tropical country but searching a collaboration from an international organization that may provide related facilities so that you may achieve some new goals with application of modern technology.Also,remember that much breakthrough in parasitic diseases has been made in tropical countries including India and not in European countries.

PART-B

6

History of Parasitic Research

It will be unjust not to describe history of Parasitology (31) and a little bit about Parasitology (chapter 8) so that our students may know the earlier works done in this field and how research has advanced by diligent observations of the eminent scientists when they were not having such sophisticated facilities.

As helminth is visible with naked eyes, it was this parasite which man encountered first in its journey of life sciences or natural sciences,as was known in ancient times. Some of these were known even before the era of biblical Palestine and ancient Egyptian civilization. Hippocrates (460-357BC), 'Father of Medicine' knew about pinworms (*Oxyuris equi)* of horses. In his "History of Animalium", Aristotle (384-322 BC)(32) has mentioned three kinds of helminthes-flatworms, cylindrical worms (*Ascaris lumbricoides*) and thinworms (*Enterobius vermicularis*). The treatises of India had the information on large roundworms of intestine (*Ascaris*), pinworms, leeches, bed bugs and mosquitoes.

The nematode, *Dracunculus medinensis* or guinea worm in common language, was known since antiquity in middle east and Africa and extruded from skin opening of humans (causing sore). Man developed a method of rewinding its extruding part on a wooden stem to remove it slowly but completely from human body. According to one school of thought, this became the emblem of modern medicine (489 BC) which everyone is witnessing while other school believes it as a serpent (not guinea worm) entwined rod wielded by the Greek god Asclepius- a deity associated with healing and medicine (presently the medical emblem features 2 serpents winding around an often winged staff–symbol of Hermes).

However, it will be erroneous to say that these ancient persons were also aware about parasitic life cycle or even bad effects of these parasites on man or animal's health. In fact, Aristotle has suggested life generates spontaneously- 'spontaneous generation' hypothesis (33); likewise, these parasites were suggested to generate from body excretions of man and animals and some considered them responsible to purify the body while many believed them beneficial to the health of man and animals and this myth about benefit of parasites persisted long after -till confirmation of germ theory of disease in mid nineteenth century.

During this ensuing period, some more parasites were discovered e.g. *Fasciola hepatica, Diphylobothrium, Strongylus, Taenia, Echinococcus.* Still we are skeptical if any one provided proof of their harmful effect on the health of animal or man. Indeed, at that time, Miasma theory of disease (34) was widely recognized which suggested that this is the pollution by the bad air that causes disease; this pollution is caused mainly by rotten organic matter (malaria was also thought due to bad air hence named - malaria). There was,perhaps, no one to believe that parasites may cause any harm to animals or man.

A new impetus to this search of parasites was given when first microscope was invented by Leeuwenhoek in late seventeenth century. Leeuwenhoek was fascinated to see microbes in water; in 1674 he described a coccidian oocyst in rabbit (*Eimeria stiedae*) and in 1681 he observed *Giardia lamblia* in his own feces (35).

Questions started erupting against spontaneous generation theory and in 1668, Italian physician Francesco Redi provided proof against spontaneous generation (finally, Louis Pasteur by his experiments proved that life can not generate from nothing and rejected spontaneous generation hypothesis). There were also questions on Miasma theory of disease and in 1700-1701, Nicolas Andry de Bois Regard (1658-1742) published his book in French in 1700 "De la generation des vers dans le corps de l'homme. De la nature et des especes de cette maladie, les moyens de s'en preserver et de la guerir "which translates "The generation of worms in human body. The different kinds and types of the disease and the ways of preventing and curing it which included his experiments with microscope. He was of the *belief* that the microorganisms which he called worms were responsible for small pox and other diseases but this was without support of any experiment (36). This work earned him the title of "Father of Parasitology".

However, an experimental proof of germ theory of disease was provided,much latter, by the Italian Entomologist Agostino Bassi (1773-1856). He demonstrated between 1808 -1813 that a 'vegetable parasite' (a fungus - *Botrytis paradoxa* now known as *Beauveria bassiana*) caused a disease in silkworms (*Calcinaccio*) (indeed, this disease was devastating silk industry of that country); this work was published in 1835 (37). In

1844, Bassi stated the *idea* that not only insects but also human diseases are caused by other living organisms e.g. measles, syphilis, plague.

A somewhat indirect support of this germ theory was encountered by the Hungarian obstetrician, Ignaz Semmelweis who, in Vienna General Hospital, noticed, in 1847, high incidence of death from puerperal fever among women who delivered their offspring in the hospital with the help of doctors (who were coming directly from autopsies) than those attended by midwives. Semmelweis (38) suspected puerperal fever as contagious and some matter from autopsies was implicated; he directed the doctors to wash their hands with chlorinated lime water,prior attending deliveries and this reduced mortality among woman from 18% to 2.2%. Nevertheless, Semmelweis and his theories were,at first, rejected by most of Viennese medical establishment.

The final proof of this germ theory of disease came with the formal experiments,conducted by the French scientist, Louis Pasteur between 1860-1864. He discovered pathology of puerperal fever, pyogenic vibrio in the blood and suggested using boric acid to kill these microorganisms; in fact, this was a land mark discovery which had long lasting effect on medical history (39). Before the French Academy of Sciences, Pasteur, on May 3,1880, delivered his famous lecture " On the extension of the germ theory to the etiology of certain common diseases". (prior to this (1860-1869) Joseph lister (1827-1912) developed and put to trial "antiseptic surgery " showing importance of microorganisms-this was as per proposal of germ theory of Pasteur). The work of German physician Robert Koch (1843-1910) and Koch's postulates (40) further strengthened this germ theory of disease which changed the scenario of medical sciences including parasitology or tropical diseases and helped in control of many communicable diseases.

Our aim of stating all above facts is to emphasize that although parasites were known since antiquity, it is most,unlikely, if they were implicated affecting health of man and animals prior emerging germ theory of disease- e.g. we may assume that prior nineteenth century,there were hardly anyone to confirm that parasites cause disease. Malaria was known with intermittent fever since long, and its treatment with cinchona bark also came in existence but etiology of malaria was considered not a protozoa but bad air,and was supported by Miasma theory. Contradicting the Miasma theory, Lanscisco in 1717 in Italy suggested that malaria is caused by animal elements. However, this animal element was confirmed much latter by Alphonse Laveran, a French army physician, in 1880 who discovered malaria parasite in human blood and described it as Laveran's bodies (obviously, his first communication on malaria parasite received with much skepticism). By this time, Pasteur germ theory also came in light and we may say it was Laveran (41) who first time incriminated a protozoa as a causative agent for the well known malaria fever; for this unique discovery he was awarded noble prize in 1907. This research followed with many other discoveries incriminating different parasites for different ailments in man and animals. Science of parasitic diseases enriched tremendously within one hundred years when many parasitic diseases, like schistosomiasis, filariasis, amoebiasis, trypanosomiasis, theileriasis, piroplasmosis, trichinellosis, paragonimiasis, fascioliasis, amphistomiasis, etc were identified along with full details of the disease, their control and treatment. A few of them have been mentioned in the outline of parasitology.

In recent years, we recognized many parasites or their stages that were prevalent in ancient times. There are reports of presence of schistosomes in mummies of 2000BC. Likewise, hookworm and other eggs have been demonstrated in the fossilized feces (42). By these accounts,we are claiming existence of parasitic diseases and their miseries since that time. However, in my opinion, it may be erroneous to consider the past situation identical to present scenario as every parasitic disease is a result of host parasite relationship, which depends on many factors and is changing as per time of host parasite associations. Therefore, on the one hand, it will be wrong to simulate parasitic ailments like of present time, on the other, it is interesting to search how parasites affected man and animals, during that period which was so different from the present one; we need to study genetic make up of host and parasite, populations dynamics, transmission factors of the two periods to reach any conclusion.

War is known to kill human beings of two sides and certainly world war one and world war two are black spots in human history. However, strange it may be that these world wars boosted research on parasitic diseases and this was the army efforts that resulted in many important discoveries on parasitic diseases which saved lives of many human beings.

NOBEL PRIZES

This history of parasitology will be incomplete if we do not mention the noble prizes won by the scientists who have worked on parasitic diseases. The first noble prize was awarded to Sir Ronald Ross in 1902 for describing life cycle of malaria parasite and how mosquito is transmitting this disease to the host. The second noble prize was awarded to Alphonse Laveran in 1907 in recognition of his work on the role played by protozoa in causing diseases; he gave half the prize to found the laboratory of tropical medicine at the Pasteur institute, Paris. The third prize, related to parasitology, was awarded in 1948 in Medicine to Paul Hermann Muller (43), born on 12th January,1899 in Switzerland,for his discovery of the highly efficient DDT (dichlorodiphenyltrichloroethane) as a contact poison against several arthropods, including mosquitoes. This synthetic insecticide was patented in 1940 and used extensively in the second world war. Obviously, its initial use completely eradicated malaria from many island areas. This compound greatly helped in Agriculture entomology also. In realty, with the discovery of DDT (and alike insecticides), man became over confident of eradicating mosquitoes and thereby eradicating malaria scar from human race. But soon man realized it was his over confidence and limited knowledge how biology is evolving itself in nature as he encountered development of DDT resistance in the mosquitoes and this award winning chemical failed to kill the mosquitoes. In fact, this is the story of many parasitic diseases where man considered it's easy to eradicate when he invented any effective drug against particular parasite but soon realized how effectively the parasite evades the lethal effect of the drug by changing its genetic makeup. Therefore, we may sum up that story of parasitology will continue and there will be a constant fight between man and his enemies-the parasites.

INDIAN PARASITOLOGY

It is prudent to mention a brief history of parasitology related to India,so that research scholars may understand Indian achievements and obligations. Though many scientists have provided details of the research work carried out in India (44), there appears two authentic review papers on history of parasitology related to India. The first (45) is by Bhatia BB and according to him, work on modern Parasitology in India may be traced back to 1869 when some British and Indian scientists started working on the subject. Griffiths Evans recovered,first time, *Trypanosoma evansi* in 1880 from horse and camels who were suffering from 'surra'. It is most unlikely if he associated this protozoa to the disease condition of the animals as credit is given to Alphonse Laveran who first time suspected a protozoa, in 1880,to be responsible for malaria (see above-41). Sir Ronald Ross made path breaking discovery at Hyderabad in 1897-1898 of finding *Anopheles* bite being responsible for transmission of malaria. Cunningham discovered *Leishmania tropica* in 1885 while Leishman (from Kolkatta) and Donovan (from Chennai) in 1903 ascribed *Leishmania donovani* as causative agent for Kala-zar ailment in man. Montgomery in 1906 discovered in Mukteswar UP, three schistosome species (*Schistosoma indicum,S.spindale,S.bomfordi)* from horse and cattle. Gaiger published in 1910 and 1915 his checklist on helminthes of domestic animals in India.

Chandler in 1926 described a new schistosome species from man from west Bengal on the basis of recovering schistosome eggs - *Schistosoma incognitum.* Rao (1933,1934) incriminated a new *Schistosoma* species (*Schistosoma nasale)* to the cause of nasal granuloma, so rampant in South Asia. Another important development in pre-independent India was publishing a series of Fauna of British India and a book in 1935 by G D Bhalerao on "Helminth parasites of the domestic animals in India (see chapter-14).

Another review paper is by Agrawal (46) who deals the subject mainly after independence of the country. In summary,the first part of twentieth century has witnessed concentration of research work on the morphology or taxonomy of the parasite, its life cycle, finding its intermediate and final host range resulting in discovering many new parasites -many were confined only to South Asian countries. When important work on these aspects were completed, scientists diverted their attention to work on epidemiology, pathogenesis, biochemical studies, immunological work including immunodiagnosis, chemotherapy on important parasitic diseases of man and animals.The latter review has dealt summary of research work being carried out in India on important parasitic infections namely coccidiosis, sarcocystosis, trypanosomiasis, theileriosis, toxoplasmosis, schistosomiasis, fascioliasis, amphistomiasis, cestodes, gastrointestinal nematodiasis, arthropods and a comment on emerging parasitic diseases.

The place of India in parasitological research in world forum is at significant level due to large research work being carried out and also its geographical spread, myriads of parasites existing in the country and a varied environment in different parts of the country giving rise spectrum of the problems. The country has made important contributions which are of great value to other Asian and African countries in dealing parasitic diseases in these countries. In recent years, India is witnessing publishing text and reference books (chapter 14) which deal the topic in more comprehensive way and have been referred by contemporary scientists to understand the diseases and related problems particularly of this region.

7

Nature the Greatest Laboratory

Our nature is the greatest or only laboratory where a large number of events are occurring at every fraction of a second - there are so many events, creatures, miseries taking place in the nature. Rightly it is considered an unparallel and biggest laboratory, man has ever witnessed. However, understanding it's any event requires deep devotion and understanding,then only success will come. Our Philosopher and Poet Kabirdas has understood these facts as have reflected in his various couplets *"Jin khoja tin paiya, gahre pani paith"* or *"chalti chaki dekh kar diya kabira roye, do patan ke beech main sabit bacha na koi"*. For an ordinary person earth is still flat and he may not know the reasons for change of seasons.

Indeed, all path breaking researches have been made by observing and understanding the events taking place in the nature. Fire was developed by friction of two stones, wheel was the result of observing running of man where legs were making a circle like structure during fast run. Agriculture also developed by observing spurting of seeds, placed in the soil. Charles Darwin was fascinated with the changes taking place in a group of pigeon and dogs in nature and he was compelled to think that it is the nature and not God who is responsible for origin of species. There was a common belief among dairymaids that they will never have small pox as they had cow pox. Even a link between malaria and mosquito was long back suspected by observing close relation of the ailment with foul air (this is the reason to name it malaria) near water bodies which were inhabiting mosquitoes. There are many examples that may be cited where man learnt by studying the event in the nature. The only difference was that they devoted time, intelligence and diligence to find the cause of the event and came out with new discoveries.

However, many times nature does not reveal itself as expected and man gets confused as he has limited power of senses which may not be sufficient to understand grossly its mysteries. For instance, our eyes have the limit to see the smallest sized matter beyond which it fails to observe; this was perhaps the reason why our great philosophers like Hippocrates failed to see smaller eggs of insects etc hence proposed life generates spontaneously. We were unable to see ultra rays by naked eyes ; likewise, our ears cannot hear ultra sounds. Therefore, such cases could be discovered only when we developed finer equipments to detect such matters. This supports the importance of extra equipments that enhance our power of observations many fold resulting in new discoveries of the nature.

In nature, it also becomes difficult to co-relate one organism with another-whether it is a part of life cycle of an organism or are two separate creatures or species,as per new terminology. This problem came in earlier studies of parasitology where scientists were confused by observing two distinct creatures of different morphology and in different host species-whether they are part of life cycle of a parasite or two different species ? For instance, scientists observed cysticercus in ruminants and tapeworm in canines- as the two parasites were living wide apart with different morphologies, obviously these were considered two different parasites as it was impossible to think them as two stages in development of a tapeworm. Later, a suspicion about one parasite with two stages started accumulating and there came the importance of experimental parasitology.

EXPERIMENTAL PARASITOLOGY

This mystery was solved by Kuchenmeister in 1850 by feeding cysticercus to the rabbits and dogs and obtaining adult tapeworms (*Taenia pisiformis*), proving that cysticercus is intermediate stage of a tapeworm (47). Actually, this is the work which firmly established importance of experimental work in understanding many mysteries existing in the nature. In fact, first experimental work in parasitology was undertaken by PC Abildgaard (48) in 1790 by feeding sexless tapeworms (ligula) to the birds and recovering mature tapeworms from them. Thus, importance of experimental parasitology or laboratory work cannot be denied in understanding basic facts of the parasite or mysteries of the nature. Indeed, these and alike works were responsible for understanding life cycle of many parasites with identification of their intermediate hosts.

In modern times, importance of laboratory work has been realized greatly as this is only laboratory work which has resulted in development of new drugs, insecticides, vaccine, antibodies, diagnostic kits and many important equipments. Such work has been extended to the extremes resulting in confining almost all research in air -conditioned laboratories and man became alien to the nature.

No doubt, development of new molecules or diagnostic kits require controlled environment hence importance of such laboratories cannot be denied. But all these researches are limited to preliminary stages- till development of the new molecule ; however, any developed molecule, a new drug, a synthetic insecticide, a vaccine molecule or diagnostic kit has ultimately to be tested under natural conditions.

Therefore, modern research requires a synchronization of both the work - laboratory and field work. It may also be in reverse direction, where we collect material from field or nature and examine them, by different ways in the laboratory. For instance, molecular epidemiology is the branch where parasitic stages are identified not on morphological grounds but on biochemical or molecular basis, being more precise and specific ; this work has to be carried out in the well developed molecular laboratories but for interpretation of the results,field data are essential and without them, it cannot be ascertained how many hybrids/strains of a particular parasite is existing with their interplay and the complexities,created by these mixing in a particular geography. Further, how this molecular interplay between hybrids/strains is influencing host susceptibilities, biological parameters, pathogenecity can only be assessed by gathering field data from the nature.

Importance of nature or field work can be explained by another example. We were working on immuno-diagnosis of schistosomiasis in our laboratory (2) where a group of goats were experimentally infected with specific number of schistosome cercariae while another group was maintained as uninfected control. Blood was collected from both the groups at given time intervals and an immunodiagnostic method namely cercarian hullen reaction (CHR) was applied on them.Our repeated results showed the diagnostic method being 99-100 % sensitive (percentage times able to detect the infection in affected animals) and 98-99 % specific (percentage times remained negative in uninfected animals).Thus it may be speculated to have similar results by applying the method in naturally infected or uninfected goats. However, the results of field cases differed from our laboratory results, diminishing the sensitivity as well as specificity of the diagnostic test. The reason for this variation is not difficult to speculate. Under laboratory conditions, we infected our goats only with schistosomes while it is not the case in nature; likewise, we are not sure what other infections are existing in these schistosome negative natural goats which are providing false positive CHR results. But one fact has emerged that there are other natural infections in the goats that influence sensitivity and specificity of CHR test.

Therefore, it is necessary to remain in touch and observe what is happening in nature with the drugs or diagnostic tests etc developed in our laboratories under controlled atmosphere. Without checking them in the field or nature, all our research is half done and its utility cannot be claimed with confirmation. Therefore, both works, undertaken in laboratory and field conditions, are equally important and complimentary to each other and one should not ignore either.

The critical analysis of above facts, clearly show that it is absurd to divide research artificially into two main categories (laboratory work and field work) as if both are miles apart having no relation with each other or giving importance of one over other. Laboratory is where our scientists are conducting research to develop new drugs or new immunodiagnostic tests, forgetting completely that ultimately this is under natural conditions where the new drugs or new diagnostic tests are finally to be tested and proved effective. There are many instances, where scientists found a new drug effective against certain pathogen in his laboratory but when the drug was tested under field conditions or in nature, the drug could not prove so effective. Likewise, in absence of laboratory work, it is not possible to develop any new drug molecule or a diagnostic kit. Therefore, it is wise to conclude that in present circumstances both are complimentary to each other and we can not deny importance of one branch of parasitology over other.

8

Parasitology and its Branches

Parasitology is the branch of biological sciences that deals with the study of parasites and all the related aspects. This leads to a basic question who is a parasite ? In fact, this term is frequently used in our society for those persons who are doing nothing and depending on others for their survival. The term Parasite means the organism (parasite) which depends on other organism (host) for its nutritional requirements. By definition, it excludes the organisms which are free living and do not depend on other organism for their food requirements. There are other details in fulfilling its nutritional requirements and important one is that in doing so, it causes some harm to the host other wise it will not be a parasite but a commensal. By this definition, all bacteria, virus, fungus, helminthes, protozoa, mites who depend on any host for its food requirements, are parasites but Parasitology confines itself to the study of the parasites which belong to animal kingdom i.e helminthes, protozoa and arthropods. Bacteria, virus and fungus are excluded in this study as these creatures belong to Plant kingdom; there is another big difference that while all bacteria,virus and fungus are unicellular, it is not the case with Parasites belonging to animal kingdom - though protozoa are unicellular, all other parasites, either helminthes or arthropods are metazoan or multi-cellular in nature. Again, the unicellular protozoa are characterized by a eukaryotic type cell (a distinct nucleus) while bacteria and fungus have Prokaryotic cell (where chromosomes are not separated in a nucleus). The hosts for these parasites may be either plants or animals; but Parasitology, taught in Medical, Veterinary and Zoological sciences, refers those parasites which are using animals as their hosts. No doubt, plant helminthes are important for well being of plants and are part of Plant Pathology or Plant Microbiology- but this is not our present topic.

The animal parasites may localize on external body (ectoparasites) or establish internally in different body systems and are called endoparasites. Helminthes, endoparasite by nature, are generally known as worms and are of two types. One is true worms or round worms or nematodes which have a well recognized alimentary canal, body cavity and are cylindrical in shape. The other group of helminthes is that of Platyhelminthes or flat worms or leaf worms which are devoid of a body cavity and where either alimentary canal is absent (cestodes) or it lacks anus (trematodes). Arthropods are characterized by jointed appendages and external skeleton. All the insects, including mosquito, house fly, cockroach, flea, ticks, mites belong to this group. They inflict injury to their host in two ways; one is direct way of sucking blood or causing disease and other indirect way is acting as vector for important viral, bacterial, parasitic diseases. In contrast, protozoa are unicellular and endoparasites, either intra-cellular or extracellular in nature.

BASIC PARASITOLOGY

This Parasitology is studied by three different faculties- each is having its own specialization. One is zoology which studies basic research on the parasite. This includes basic aspects like morphology which may be gross, microscopic or by electron microscopy, evolution, life cycle, physiology, biochemistry, factors responsible for survival of parasite in nature and its spread, development of species,subspecies and strains, its intermediate hosts, interactions between parasites and intermediate hosts and with the environment. An important aspect of these studies is taxonomic studies where the particular parasite is assigned its right place in animal kingdom and to identify specific characters which help in identifying family, genus,species and sub species of the parasite. Identifying new species of the parasites is an important research aspect in basic sciences. The basic research also includes developing genomes for important parasites; so far, full genome is known for some important parasites like *Schistosoma* sp, *Nicator* or hookworm, *Plasmodium* sp (49) and work is continuing to map genome of many other important parasites.

APPLIED PARASITOLOGY

Medical Parasitology is related to parasites of humans while Veterinary Parasitology deals with parasites of domestic animals and poultry - even then, there remains a spectrum of animals belonging to mammals, reptiles, birds, fishes, amphibians and parasites inhabiting these large number of animals are dealt by the zoologists who are having interest in parasitology research.

Medical Parasitology and Veterinary Parasitology are considered applied branches of parasitology as they deal mainly diagnosis, treatment and control of parasites responsible for causing diseases to the man or his animals. It is unfortunate that till recent years, medical parasitology was synonym to Malaria in India and almost all attention was paid to malaria,ignoring other important human parasites. No doubt, malaria is an important protozoan ailment and needs greater attention, but in the tropical country like India, there are other parasites also that are affecting large rural population and await their solutions. Following is a short description of other parasitic diseases,affecting human population.

MEDICAL PARASITOLOGY

The guinea worm (*Dracunculus medinensis*), *Ascaris, Trichuris* -the nematodes by character, may not be endangering human life but man come in contact with these helminthes which are vast in their spread, influencing more than one continent and causes great discomfort to human beings. Then, there are other helminthes which are causing loss of human life. Among them, mention may be made of blood flukes or *Bilharzia* or schistosomes. They are called blood-flukes for their presence in the blood vascular system of the host, *Bilharzia* as it was first discovered by the German Surgeon Theoder Bilharz and schistosomes for having a splitted body. The helminth causes urinary and hepatic Schistosomiasis in man and animals when they acquire its larvae while wading in the polluted waters. These are the fresh water snails like *Bulinus, Biomphalaria* and *Oncomelania* sp which are responsible for their transmission. According to an old record, 200 million people are suffering from schistosomiasis while 600 million are at risk in the endemic areas. The parasite is existing globally but is a main problem in African,Asian and Latin American countries where it is considered only next to malaria in transmitting the damages. The parasite spreads in new geographical areas as humans attempt to increase irrigated lands for enhancing agriculture produce. Existence of human schistosomiasis in India has been a controversial subject,although cercarial dermatitis is rampant in Indian rural population (2).

Filariasis (*Wuchereria, Burgia* sp) or elephantiasis is disabling humans by causing permanent physical deformities which,if once developed, difficult to cure by medication. The helminth spreads through a mosquito of *Culex* sp and more common in hilly tracts. Filariasis or elephantiasis is still causing physical disability to poor community making them unfit for employment hence these poor patients can be seen taking shelter in many religious places of our country. There is another important filarial parasite, i.e. *Onchocerca volvulus* which is occurring in African countries and transmitted to man by *Simulium* flies, commonly known as black fly. The adult filariid lodges itself in the skin leading to nodule formation but main damage is by migrating its microfilariae to the eye causing sight deformities even leading to blindness, when concentration of microfilariae reaches to its height. This disease is commonly known as River valley blindness as larvae of its vector develop in the river. When rate of blindness in a village becomes significant, the fearful young people abandon their homes thereby affecting farm production and economy of the village. Fortunately, its vector species is confined to African continent, hence the problem is limited to that geography alone.

Hookworms or *Ancylostomum / Nicetor* sp is another nematode affecting rural population and is an important factor for creating anemia in children and women-endangering life of pregnant ladies- particularly in rural areas.It is a soil helminth and does not require any intermediate host for its development and transmission hence existing in greater part of the geography where environment is favorable for survival of larvae in the nature. Whenever people travel bare footed, the larvae penetrate host's skin to develop to maturity in the intestine where it not only sucks blood but causes ulcers in the intestinal wall, too. So is *Toxocara canis,* whose normal host is a dog but its infective eggs remain in contaminated soil which is swallowed by children during play ; the larvae are liberated from eggs in the intestine and migrate to different body parts causing visceral larva migrans where infection to eye is an important complication and at times has led to extrusion of eye by suspecting tumor growth.

Hydatidosis is the larval stage of dog cestode (*Echinococus granulosus*) which man acquires by ingesting its eggs,present in the contaminated soil. The cyst develops mainly in abdomen and lungs. But cysticercosis develops in man by acquiring *Taenia solium* infection by eating infected pork. Cysts are more dangerous when develop in brain leading to severe pathology including epilepsy. Importantly, former infection spreads through street dogs by soil contamination with the eggs while latter spreads in man by pork eating and in pigs by man contaminating soil with his feces excreting eggs of the tapeworm.

Above are helminth infections which affect large number of countries and continents while there are some other helminthes that may be important only to some countries. Mention may be made about trichinellosis (*Trichinella spiralis*) which occurs in man by eating infected meat of wild animals like boar and bear ; lung fluke (*Paragonimus westermani*) infection which occurs by eating infected crabs and fishes - as it affects lungs of man, the symptoms are just like tuberculosis though treatment is quite different from this ailment. A type of hydatidosis is caused by *Echinococus multilecularis* where the cysts behave like malignant cancer and is capable to develop in other parts, if detached from its original location-obviously, its infection proves fatal.

Among protozoan infections, everyone is well aware about Malaria which is caused by *Plasmodium* sp and spreads by biting *Anopheles* female mosquitoes. It is most important human infection where research is being continued since last many years but still we do not have any vaccine to control malaria while mosquitoes have developed resistance against insecticides.

A somewhat equally important group of protozoa is trypanosomes whose infection also causes great havoc to human population. There are different groups of this protozoa and certain trypanosome species (e.i. *Trypanosoma gambiense, T.cruzi*) are causing important human diseases commonly known as sleeping sickness, chagas disease but, are confined to African and south American countries as its intermediate host (Tsetse fly,triatomid bug) is confined to these geographies. However, its presence in Africa has adversely affected human dwellings and animal husbandry in certain pockets despite a very favorable climate for Agriculture. Again, its sibling namely *Leishmania donovani* is more wide spread including India as this protozoa spreads by sand flies.The disease caused is commonly known as Kalazaar in India and wide spread in Eastern parts of the country, responsible for heavy mortality in children.

There are other protozoan diseases which are important to certain category of persons; amoebiasis is a problem where water contamination with fecal material is more common and may prove a great public health problem in certain places; toxoplasmosis is one which may not affect adversely pregnant women (though may cause abortions) but it may cripple the new born,affecting their nervous system. There are certain protozoa like *Cryptosporidium* and *Pneumocystis* which are gaining importance as they are causing secondary infections in HIV infected human beings and responsible for complications including death. Irony it may be that India, with so much suffering from parasitic diseases, do not have separate department of Parasitology in our state medical colleges.

The study of Medical Parasitology does not confine to know the parasite, its morphology or identification but there are many other aspects that are studied -in that sense Parasitology is a multidisciplinary subject. The first question will be how the parasite enters in human body and its mode of transmission, methods of existence in nature (epidemiology), pathogenesis and pathology including biochemical and immuno pathology. An important aspect is diagnosis of the disease which has two components - one is clinical diagnosis which depends on studying symptoms and identifying the disease. The other is laboratory diagnosis which may be parasitic or molecular i.e. biochemical, immunological diagnosis. In parasitic diagnosis, parasitic stages like eggs, larvae are detected but it is not a sensitive way as many times, parasitic stages are not detected. Therefore,, other way of diagnosis i.e. immuno-diagnosis is followed where either specific antibody or parasitic antigen is detected. The most important aspect of Medical Parasitology is treatment which is co-related with cure of the patient. However unlike to bacterial infections, there are not many drugs against parasitic infections; in fact, there are certain parasitic diseases which do not have specific medication and surgery is more effective option e.g. hydatidosis. Though, control of parasitic diseases (community medicine) is another aspect, it has not been attended seriously and has been confined to chemotherapy. Still there are no effective vaccines that may be used in controlling parasitic diseases leading to a great impediment in preventing the infections. Thus absence of serious control measures and changes in the environment and life pattern of human population has resulted in increase in prevalence rate of many parasitic diseases creating health problems.

There is a specific category of parasites, both helminthes and protozoa which are capable of affecting both animals and man - this has led to a separate discipline of Parasitology and is called Parasitic zoonosis and studied by all groups of the Parasitologists.

VETERINARY PARASITOLOGY

Veterinary Parasitology deals a large spectrum of parasites those affect our domestic animals and poultry. These domestic animals include cattle, buffalo, sheep, goat, camel, horse, ass, donkey, pig, dog and poultry. With changing of time, studies of parasites of wild animals, Yalk, Mithun and birds like turkey, esmeeo, *bater* (guinea fowl), have been included specially in specialized courses of Veterinary Parasitology. The study of parasite in each animal species includes morphology of the parasite, life cycle, epizootiology, pathology, diagnosis, chemotherapy and control measures. Obviously, this causes a large syllabus and scientist generally selects only a limited aspect for his research studies.

Name of important parasites of veterinary field differ from that of human parasitology. As animals are allowed for grazing in pasture lands with free assess to water bodies, flukes are important parasites which are present in these environment and affecting animals. Among fluke infections, *Fasciola, Schistosoma* and amphistome species are important in causing production losses as well as mortality in young animals. All these flukes spread by fresh water snails which are present in different water bodies. Among nematodes, *Haemonchus* sp is most important particularly in causing heavy sheep mortality with increasing problem due to development of drug resistance. Hookworms (responsible for blood loss),

Oesophagostomum (produces nodules in the intestine) are also important in veterinary Parasitology.

Among protozoan infections, trypanosomes are important as they cause great pathology in animals in African countries making animal husbandry difficult in certain parts of the continent. These trypanosomes are not prevalent in other countries due to absence of particular vector (Tsetse fly) but an equally important trypanosome is *Trypanosoma evansi* which spreads through a horse fly (*Tabanus* sp) and widely prevalent in Indian continent, middle Asian countries, and is responsible for a chronic disease (surra) in domestic animals. Theileriasis, caused by *Theileria* sp and spread through hard ticks, is gaining greater importance in cattle cross breeding program where cross bred cattle succumb easily because of theileriasis, thereby jeopardizing the whole cross breeding program.

Coccicidia and coccidiosis has proved a dangerous proposition to whole poultry industry as it affects young chicks very severely wiping whole flocks within a short time ; there was a time when survival of whole poultry industry was in danger due to coccidiosis. Coccidia and coccidiosis affects not only poultry but also young animals. The main reason for the problem is that coccidian life cycle does not require any intermediate host and its oocysts,present in the droppings hence soil, become infective within three -four days under suitable conditions of humidity and temperature.

It is true that less attention is paid on Veterinary Parasitology considering that parasites are not causing death of the animals and leading only production losses. However, presence of effective vaccines and antibiotics against bacterial and viral infections have now diverted our attention to parasitic diseases. Further, the recent researches have confirmed that parasites are not only responsible for production losses but may cause heavy mortality wiping out even whole flock if correct diagnosis and treatment is not carried out within stipulated time. This has compelled the authorities to change the complacent view on parasitic diseases and pay more attention on them to check both production loss and mortality.

9

Research Scope in Parasitology

A close look on three disciplines of Parasitology reveal that these are the branches where a scholar initially select them as per his basic qualifications. But while pursuing research in Parasitology, the line of demarcation becomes blurred and it is in fact combination of a group of scientists,from varied disciplines, who are able to solve any one problem; at that point, it is absurd to divide the subject as medical or veterinary or basic parasitology. In fact, these will be your interests and experiences which will decide your line of research and will determine your achievements.

Additionally, twenty first century is confronting with new problems which are the results of our past experiences in dealing parasitic diseases. Earlier, man became confident of eradicating parasitic diseases completely from the continents as soon as he developed specific drugs with high efficacy for killing the parasite and use of insecticides for eradicating flies. But soon, man realized that parasites and arthropods are smarter than considered earlier as they know better how to escape these adverse effects on their life and to survive with new vigor. An alternative method of controlling parasites is biological control but this is also marked with its own problems while use of insecticides, beside resistance among insects, is polluting our environment. Many parasites have been incriminated to cause benign or malignant tumor or autoimmunity and research is continuing to unravel these mysteries. Pathogenesis of some parasitic diseases is re-worked to confirm whether associated bacteria or virus is responsible for severity or the parasite itself ; it is becoming interesting to study how immunodeficiency cases like organ transplants, HIV infection are affecting susceptibility and pathogenicity of the parasitic infections. More attention is being paid to know why some persons or animals in a community either remain resistant to a parasitic infection or do not show severity of the infection and remain a career for whole life.

It is generally lamented that Indian scientists are doing a repeat research of what so ever is carried out in western countries and there is nothing new in Indian research work. This is one reason why our research papers do not find place in western journals and our Indian journals have a low impact value. To some extend the blame appears true. If we review the past research achievements in Parasitology, almost all new concepts, - zoo prophylaxis, natural nidus, zoo geology, heterologous or homologous immunity, vector immunity, autoimmunity, self control phenomenon, post parturition nematode rise, tick immunity, *in-vitro* cultivation of parasites, parasite antigens in excreta, phylogenic studies of the parasites etc have emerged from abroad and Indian scientists merely re-confirmed the findings or described them in different situation or host. Western dominance in the science may also be visualized by the publication of reference books, number of parasitological journals, advances in Parasitology all coming from foreign countries ; till recently, text books on Parasitology were also imported from western countries and its only in recent years that India have started publishing books on Parasitology including some reference books which are being consulted not only in India but other countries,too.

There are various reasons for this dominance of American and European countries in science including Parasitology though tropical countries (mostly Asian and African countries), should have their upper hand due to their geographical location, environment and endemicity of the infections, hence discovering new findings related to parasitic diseases. No doubt, many parasitic diseases are not occurring in European countries or in USA but scientists of these countries are dominated in international organizations and by their seer resources and existing facilities, they are able to take up epidemiological and other related research in endemic countries thereby filling the gap of non-existence of the diseases in their countries. Such epidemiological studies are only a part of their research program, else they are devoting their time in understanding fundamentals of the parasites -it may be related to genes of the parasite or developing a mosquito resistant to malaria parasite or trying to develop a parasite *in vitro* or vaccines and diagnostic molecules for important parasites. This is all because of man-power, funds, equipments,library facilities which are not up to that mark in endemic countries.

However, this blame of copy cat the research work of western countries may to some extend be erased if our Indian Parasitologists instead of taking copy cat research projects may start concentrating on local problems with visualizing the changes in nature, environment, demography, animal husbandry practices and parasite population. Certainly, such original work will lead to new findings which will get important place in scientific literature.

I am herewith reproducing my two edited lectures, delivered in December, 2013 in the winter school on "Molecular biological approaches for diagnosis and control of parasitic diseases" at IVRI, Izatnagar,UP merely as a thought provoking exercise.The first lecture highlights restricted importance of immunodiagnostic tests in schistosomiasis hence raises questions giving great priority on the subject ignoring other important topics of the infection. The other lecture discusses how difficult it is to control any parasitic disease,with schistosomiasis as an example, when we do not possess required knowledge regarding that disease. Hopefully, study of these lectures will force our research scholars to select India's related problems or those having some impact on development of parasitology.

A DISCUSSION ON DIAGNOSIS OF ANIMAL SCHISTOSOMIASIS IN SOUTH ASIA

M C AGRAWAL

Former National Fellow and Emeritus Scientist, College of Veterinary Science and A H, Jabalpur 482001

Those who are working on diagnosis of schistosomiasis are aware how vast is the subject and its impossible for anyone to cover the topic in one hour lecture. Therefore, it will not be my goal to cover the whole subject, rather I will discuss some pertinent points on the topic at this occasion - the detailed discussion is given in my recently launched book (Agrawal, 2012). One advice to scientists who are interested to work on any problem of schistosomiasis in south Asia is to study the concerned literature as such exercise will avoid duplication of work and will help in understanding problems related to south Asian schistosomes and schistosomiasis which are certainly different from African or East Asian schistosomiasis.

Generally we are working and discussing methodology of one or other diagnostic technique, its sensitivity and specificity, but there are no discussions whether we need to devote our time, energy and funds for developing that specific diagnostic technique and what will be its utility in present scenario. I think such discussion is necessary and my present lecture will focus on this topic taking schistosomiasis as an example.

Historically, diagnosis was restricted to what we call clinical diagnosis but with passage of time it has gone in the back ground while laboratory diagnosis has taken a central stage-so is the case with schistosomiasis. One reason for this is our realization of difficulty in identifying the disease on symptoms alone due to commonness of them in many other ailments. Therefore, we cannot whisk away laboratory diagnosis and in fact this is the main topic while discussing schistosomiasis diagnosis. To this, I have taken liberty to include identifying schistosome infection in snail host,as well to understand the subject.

OUR AIM OF DIAGNOSIS

The basic question is what is our aim of identifying schistosome infection ? Whether it is only for treating a particular animal coming to veterinary hospital ? Or to declare a geographical area endemic with a given schistosome species ? Or to understand epizootiology of the infection ? Or to study zoonotic aspect of a schistosome species ? Or to know percentage of infected animals in each species with its age and sex ? Or to study immune status of animals after vaccination program (so far, no commercial vaccine is used against schistosomiasis in India ; it may only be on experimental basis). Or we wish to collect schistosomes during necropsy or in experimental animals for molecular and other studies? Again schistosomiasis is existing in animals in two important forms (nasal schistosomiasis or snoring disease and hepato-intestinal schistosomiasis) and diagnosis of both are different. So, there are varied aims of diagnosis hence our methodology of diagnosis should also differ as per our aim. Therefore, I will discuss the subject as per our aims and what best methodology we should follow to attain our aims of studies.

SENSITIVITY AND SPECIFICITY OF TESTS

While starting the discussion, let me clarify that diagnosis of schistosomiasis is difficult due to two reasons. First, the symptoms in hepatic form like diarrhea, dysentery, anorexia, weakness, bottle jaw etc are also present in other chronic or sub-acute infections. But in snoring disease, the animal exhibits path ognomic signs like snoring sound, nasal discharge and nasal granuloma; though totally depending on these symptoms we may miss the infection in buffaloes or some cattle which are passing eggs in nasal discharge but without symptoms; and by this strategy, you cannot identify nasal schistosomiasis,as is existing in Jabalpur, where local cattle are resistant to *Schistosoma nasale* but buffaloes are harboring the parasite- this was confirmed by Banerjee and Agrawal (1991).

Therefore, for confirming the infection, we have to depend on laboratory diagnosis which is of two types -Parasitological and immunological- and both are having their own problems. Basically, laboratory diagnosis is difficult due to poor sensitivity with no ante-mortem parasitological method being capable to correctly estimate schistosome prevalence in a given geographical area. For these reasons, I have argued about under estimation of schistosomiasis in whole of south Asia (Agrawal, 1999).

PROBLEMS IN PARASITOLOGICAL METHODS

In parasitological diagnostic method, we can deal only with feces (in hepatic form) of the animal during ante-mortem period. The most common method of fecal examination, followed in any veterinary diagnostic laboratory, is direct wet smear method. In some advance laboratories sugar flotation or salt concentration method is also followed. Ideally, both supernatant and sediment should be examined for checking nematode and fluke infections but many times only supernatant is examined without caring for the sediment. Ironically, examination of feces by both these methods have resulted in identifying only 2-8% schistosomiasis in any animal species whereas,in reality, the infection rate was varying between 60-80%. Obviously, such diagnostic method cannot be recommended for diagnosing schistosomiasis when surveillance of the infection is our aim.

ANTEMORTEM DIAGNOSIS

Therefore, my laboratory assessed different coprological methods which may be used preferentially under field conditions. We have tried and compared direct smear, salt floatation, acid-ether or formal-ether, sieving method, and Kato-katz methods in experimentally infected laboratory animals as well as in field animals (Agrawal 2000, Gupta 2002, Vohra, 2005). The results varied as per animal species, age of animal and intensity of the infection but sensitivity of these tests ascended from direct smear to salt floatation, kato-katz, formal ether methods leading highest level to 16%.

As any of the above egg detection method could not provide a practical solution of diagnostic problem, we tried hatching method by taking advantage that the eggs of schistosome species contain fully developed miracidia, ready to hatch,soon after coming in contact with fresh water. There are many factors which influence this hatching of eggs and have been discussed elsewhere (Agrawal, 2012).

In summary, we can say that hatching method proved always superior over any egg detection method hence this is the test which we recommend to be followed in field as well as in research laboratories. When we studied the method more meticulously, it was found more effective than any egg detection method where number of schistosome eggs in the fecal material was low- this is the young or old animals where egg concentration is low. Contrastingly, the hatching method, though showed higher prevalence in pigs in comparison to egg detection methods, could not show significant variation in prevalence rate of porcine schistosomiasis at Jabalpur ; it is because of higher fecal concentration of schistosome eggs in pigs (in comparison to other animal species) resulting its identification by other egg detection methods, as well.

The other animal species where we observed different results, while comparing miracidia detection *vis a vis* egg detection, was experimentally infected albino mouse. Here, acid-ether method was more simple and sensitive than hatching method which failed many times to detect miracidia in the fecal material.

The hatching method is not only simple but also cost effective as it does not require any chemical or use of micro slides. Therefore, this is an ideal method that can be followed under field conditions with little efforts.

POST MORTEM METHODS

Even diagnosing of schistosomiasis is not easy while conducting post mortem of an animal in comparison to other helminth infections. Thus a simple cut in the liver may yield liver flukes, opening of rumen reveals amphistomes while intestine lumen will cause recovery of many nematode species. And in all these organs, you will not get schistosomes declaring the animal negative for schistosome infection. I may substantiate my statement by one reference of Rathor (1998) who compiled postmortem reports of more than eighteen thousand buffaloes of 28 livestock farms of India; the buffaloes were harboring *Fasciola* and amphistome infections but no case of schistosomiasis was recorded in these post mortem reports. As intermediate hosts of amphistomes and schistosomes are identical,it is difficult to believe that all the animals were negative for schistosomes-rather this is a case of failure of diagnosing schistosomiasis in any of these animals due to following wrong post mortem techniques.

This issue was attended by us and we have described specific techniques which should be followed during conducting post mortem of an animal for diagnosing schistosomiasis in them. These are two techniques - one is perfusion technique for recovering alive blood flukes and the other is egg detection method using intestinal scrapings or liver pieces (Agrawal 2012, Vohra 2005).

IMMUNO-DIAGNOSIS

A lot of research work globally is being carried out on immuno-diagnosis mainly to study human schistosomiasis in endemic countries and there are various reviews over the subject. It is difficult to cover them in one hour lecture. Nevertheless, I will mention some important facts which will help our discussion. For more details, you may read chapter 8 of the above referred book (Agrawal, 2012).

ANTIBODY DETECTION METHODS

Initially, all the efforts were made to detect specific schistosome antibodies, developed by the host due to schistosome pre- infection by presenting different types of the antigens.

There was fascination to use live larval stages of the blood flukes,all three forms-viable eggs, miracidia, and cercariae- as experimental works have shown development of precipitates or blebs around viable schistosome eggs or immobilization of miracidia or development of a hyaline membrane around cercariae when these larvae were immersed in the serum containing schistosome antibodies. The former two larvae were discarded as the tests were not very sensitive while CHR (cercarian hullen reaction) proved a promising immunodiagnostic test though with the requirement of procuring alive cercariae for conducting the immunodiagnostic test.

To avoid the requirement of alive larvae, soluble antigens were developed using different methods and different stages of the schistosomes. Those who dealt the subject realized how difficult it is to recover a good quantity of schistosomes comparatively due to its smaller size and its location in the blood vessels. Therefore, soon, the immuno-diagnostic tests requiring larger quantity of antigens i.e. ring precipitation test were discarded. And in recent years, all attention is diverted on ELISA ; as plate ELISA requires larger quantity of antigen and is more tedious, Dot ELISA is the method of choice among schistosomologists as each test requires only nano gram of antigens and is more simple to perform.

Looking to importance of diagnosis of fluke infections,a NATP project was run by ICAR with main center at IVRI during 2000-2004 and making Jabalpur, Hisar, as investigating centers for diagnosis of schistosomiasis ; some important work was also carried out at Bangalore veterinary college. Though these centers were able to develop Dot ELISA tests for diagnosing schistosomiasis with higher sensitivity, specificity could not reach to higher levels. Even some scientists (Sumanth et al. 2004) found Dot ELISA positive in all the field animals who ever have been exposed to schistosome infections (?) and presently were either negative or positive for schistosome eggs. What will be the utility of such diagnostic test in studying schistosomiasis remains debatable (Agrawal, 2012).

ANTIGEN DETECTION METHOD

The antibody detection methods have now been discarded globally by scientific community due to four reasons. First, it does not differentiate between present and past infection. Second, it fails to calibrate intensity of the infection ; neither it reflects that all these positive animals are immune to re-infection. Third, it is difficult to collect blood from large animals under field conditions where controlling of the animals is a great problem (now there are ethical questions for collecting human blood merely for surveillance work). Fourth,the invasive method (collecting blood) merely for diagnosis is not favored due to the chances of transmitting viral and other infections; this has become more controversial in human cases where there are chances of transmitting HIV due to minor negligence of the technician.

Therefore, in recent years, attention has been diverted to detect schistosome antigen in feces or urine of man and animals and principles of ELISA are followed for developing these methods. To make diagnostic test more specific and sensitive monoclonal antibodies have been used to attach them on nitrocellulose strip. Again, single step diagnostic kits have been developed abroad where a solution is to pour on the strip which will develop two bands in case of positivity and only control band if the case is negative for schistosome infection. One single step dipstick kit from European Veterinary Laboratory, Netherland was tried by us resulting in five positive human cases who were either suffering or were having history of cercarial dermatitis. Let me clarify that no commercial single step diagnostic kit is developed or available in India, though needed for different reasons.

SCHISTOSOMA SPECIFIC TESTS

An important clarification, regarding all immuno-diagnostic methods is that they are capable to diagnose *Schistosoma* infection- only up-to genus level and none of these tests are species specific. Therefore, it is important to note that none of these tests will inform which schistosome species is present in positive host species. When it fails to identify schistosome species, any of these tests are of no use in identifying strains or sub-species of the schistosomes; neither they may help in detecting any new schistosome species.

Not only immuno-diagnostic methods, but our egg detection methods may also fail to differentiate among schistosome species where egg morphology is closely resembling. If you study the literature, you will notice that all oval shaped schistosome eggs could not be ascribed to *S.haematobium* or *S.indicum* ; likewise, spindle shaped eggs do not mean presence of only *S.spindale* in the area (Rollinson and Southgate 1987).

Therefore, now we have come to our main discussion -what is the aim of our diagnosis ?

It would have been clear from above discussion that immuno-diagnostic tests will not solve all our problems related to schistosomiasis and we have to stitch specific methods of diagnosis as per our requirements. I am discussing some of them in brief :

DIAGNOSIS FOR TREATMENT

When an animal with certain symptoms visit to a veterinary hospital, the main aim is the treatment of that particular animal. As the symptoms of schistosomiasis are common with other infections (though treatment is not same) the differential diagnosis for schistosomiasis is important. The parasitological diagnostic tests have been termed as 'Golden tests' hence it is always recommended to diagnose these animals by coprological methods. This should include acid-ether or formal -ether method along with hatching method. The tests may be repeated twice or thrice if one time method shows negativity.

COLLECTION OF SCHISTOSOMES

If there is the need of collecting alive schistosomes from any animal species, perfusion method is the ideal one (Agrawal, 2012). Recovery of schistosomes will also confirm presence of the infection in that particular host species. In old days, the scientists teased the blood vessels for checking the animal positive for schistosomiasis or for collecting alive blood flukes - this method is tedious and less sensitive hence not recommended. Previously, it was difficult to apply perfusion technique being carried out by the use of a peristaltic pump which was costly and not available in Indian laboratories ; however, we have replaced it,successfully, with a vertical water pump, most commonly used in desert coolers in Northern India (costing Rs. 500) hence now it is feasible by any laboratory to perform perfusion technique for recovering schistosomes during necropsy of the animals and details may be found elsewhere (Agrawal, 2012).

The alternative method, though less sensitive, is to collect mesentery or nasal cavity (when working on nasal schistosomiasis) of the suspected animal, cut into small pieces and soak in Luke warm saline for 4-6 hours.

Afterwards, the saline is filtered using a muslin cloth which is transferred to a Petri-dish to collect the blood flukes (Agrawal, 2012).

IDENTIFYING NEW ENDEMIC AREA

As you have come from different places of the country, it is interesting for you to confirm whether a geographical area of your state or district or Tahseel is positive for any schistosome species. For confirming this possibility, you have to examine both intermediate and final hosts- in all seasons and in good number else your results will not be reliable.

Till now only *Indoplanorbis exustus* and *Lymnaea luteola* have been identified as intermediate hosts for all the five well studied schistosome species i.e. *Schistosoma indicum,S.spindale,S.nasale,S.incognitum* and *Orientobilharzia dattai* while the snail hosts of *Bivitellobilharzia nairi, O.bomfordi* (?), *S.haematobium* or *S.gimvicum* (- Gimvi infection) have still not been confirmed. Therefore, all the fresh water snails of the area should be examined meticulously for shedding of brevifurcated, apharyngeal, non-ocillate cercariae ; presence of such type cercariae will indicate prevalence of any mammalian schistosome species in the area (Agrawal, 2012). If these cercariae are shed by a new snail species, there are all chances that you are dealing with a new schistosome species.

As morphological studies of these cercariae cannot differentiate different mammalian schistosome species conclusively,neither strains or sub-species of the schistosomes, it is recommended to develop molecular techniques to study these cercariae. So far, it is believed that the snails are infected only with mono-species of fluke infections and molecular studies will confirm or otherwise this notion.

When you are examining final hosts for schistosome infection, first confirm that the animal is not a migratory animal otherwise origin of the infection will remain questionable. Area of birth of the animal is important in such studies and only by this methodology we suspected that Jabalpur is also harboring *Schistosoma nasale* in the buffaloes (Banerjee and Agrawal 1991).

Though immuno-diagnostic methods are having high sensitivity (80-95%), their specificity ranks between 60-70% therefore solely depending on immuno-diagnosis while confirming a new endemic area is not a very good idea. However, use of only parasitological techniques may not reveal schistosome infection in a geographical area hence use of immune-diagnosis is important while exploring possibility of schistosomiasis in any animal species though it must be associated with sensitive parasitological methods which should always include hatching method. While examining animals for *S.nasale,*instead of nasal discharge, check the nasal scrapings as the latter is four times more sensitive than the former method.

STUDYING ZOONOTIC POTENTIALS OF SCHISTOSOME SPECIES

This is a topic of great interest in whole of South Asia but has not been studied meticulously. Identifying an endemic focus of urinary schistosomiasis in Gimvi village of Ratnagiri district, Maharashtra state (Gadgil and Shah, 1952) generated considerable interest in human schistosomiasis in the country but non-detection of clinical urinary cases from other parts of the country created a complacent view of absence of human schistosomiasis in other parts of the country. This view was developed ignoring the fact that *S.incognitum* was first detected in two human cases by Chandler (1926) and there are reports of finding schistosome eggs in human excreta in other parts of the country (Agrawal, 2012). The topic has been discussed in more details while reviewing present status of schistosomiasis in India (Agrawal, 2005). Prevalence of cercarial dermatitis in rural India is very common (Agrawal, 2012) although no further studies have been carried out on the ailment. Thus we are not sure to what stage of development the schistosomula reaches in humans suffering from different immunological conditions.

In the above cases, parasitological methods have failed to detect schistosome eggs in human excreta, therefore, it is advocated to use a more sensitive and specific antigen detection test. The specificity of single step diagnostic kits, developed by using *S.mansoni* or *S.haematobium,* in other countries is not very high warranting developing such kits using Indian schistosome species. Likewise, we do not have any specific immunological test for confirming cercarial dermatitis cases ; this is only history of dermatitis and recovering schistosome cercariae from the fresh water snails which is collaborated with the ailment. Therefore, it will be beneficial to develop an immunodiagnostic test which may confirm cercarial dermatitis- ideally with possibility of differentiating cercarial dermatitis caused by avian schistosome cercariae and that by mammalian schistosome cercariae.

As you will notice that immunodiagnostic test for cercarial dermatitis will, most probably, not identify schistosome species or strains involved, hence there is the need to develop molecular techniques for identifying species or hybrids or strains of schistosomes, existing in whole of south Asia. Obviously, both the methodologies are supplementary to each other in solving the problem of schistosomiasis in the region.

CHECKING HOST IMMUNITY AGAINST SCHISTOSOMES

In viral and bacterial infections, immunological tests have also been used to check immune status of a given population and are good tools for vaccination programs. When we look the scenario with relation to schistosomiasis, two facts have emerged.

First, there is no vaccine which is being used for controlling schistosomiasis. This is the case both for human schistosomiasis as well as for animal schistosomiasis. I am not aware if any institute in India is trying to develop a feasible vaccine for schistosome infection in any host species.

Secondly, our immuno-diagnostic tests are able to reveal presence of antibodies in positive cases but whether these particular antibodies will be able to protect the host against re-infection with schistosomes is questionable.So far, we have not developed any immuno-diagnostic test whose positivity also guarantee protection of the host against re-infection.

SURVEILLANCE OF ANIMALS FOR SCHISTOSOMIASIS

Whenever, we undertake surveillance work for identifying any infection, our diagnostic test should be most sensitive and specific in diagnosing that infection. Ironically, no diagnostic test stands on above criteria with regards to schistosomiasis. With Indian schistosomes, parasitological tests are least sensitive due to comparatively lower egg excretion hence their sole use in surveillance work is never advocated. In our opinion, hatching method has proved most sensitive parasitological method during ante-mortem hence must be included in all surveillance works. Those cases, positive by hatching method, must also be examined by a more sensitive egg detection method which will help in identifying schistosome species of that area.

Since parasitological tests are least sensitive, the surveillance work must be carried out by supplementing them with a more sensitive and specific immuno-diagnostic method. All the work, carried out in India, have proved Dot ELISA as most promising immuno-diagnostic test hence this should be incorporated in surveillance work but fully understanding its limitations (Agrawal, 2012).

The antigen detection test in animal feces is more simple and able to detect existing schistosome positive cases hence there is the need to develop these tests for understanding schistosomiasis in South Asia.

Our above discussion show the limitations of immuno-diagnostic tests in understanding schistosomes and schistosomiasis in whole of south

Asia without developing more sophisticated molecular techniques to identify new schistosome species and its hybrids that have developed in the continent due to hybridizations with existence of two or more schistosome species in a single host species. Thus it negates general conception that a immune-diagnostic test will solve almost all problems related to schistosomiasis.

SUMMARY

The symptoms in hepatic schistosomiasis are vague hence laboratory diagnosis is carried out for confirmation of the disease. While deciding to employ diagnostic methodology, it is important to consider our aim of diagnosing the infection. When our sole aim is treating a sick animal, visiting a Veterinary Hospital, coprological diagnosis is recommended but using more sensitive methods like acid-ether and hatching methods. During post-mortem,either for collecting blood -flukes or for confirming the infection, perfusion method is the choice having highest sensitivity. In deciding zoonotic potentials of a given schistosome species or for surveillance of the infection both parasitological and immunological methods should be followed as both will supplement each other. As antibody detection method does not differentiate present and past infection with additional disadvantage of invasiveness, antigen detection methods in host's excreta are gaining priorities. But it must be remembered that so far all immunodiagnostic methods are efficient only to identify the infection up-to genus level (*Schistosoma*) and cannot differentiate the infection up-to species levels. Therefore, it is essential to undertake molecular studies, if you wish to decide existence of schistosome species, sub-species or hybrids, in a given geographical area, as no immuno-diagnostic method can reveal such complexities of schistosomiasis.

REFERENCES

Agrawal, M.C. 1999. Schistosomosis : an underestimated problem in animals in South Asia. *World Animal Review* 92 : 55-57.

Agrawal, M.C. 2000. Final report on National Fellow project "Studies on strain identification,epidemiology,diagnosis,chemotherapy and zoonotic potentials of Indian schistosomes. ICAR, New Delhi.

Agrawal,M.C. 2005. Present status of schistosomosis in India. *Proc Nat Acad Sc, India.* 75 (B),Special issue : 184-196.

Agrawal MC. 2012. Schistosomes and schistosomiasis in South Asia. Springer India Pvt Ltd, New Delhi.

Banerjee PS and Agrawal MC. 1991. Prevalence of Schistosoma nasale Rao 1933 at Jabalpur. *Indian J Anim Sci* 61 : 789-791.

Chandler AC. 1926. A new schistosome infection of man with note on other human fluke infections in India. *Indian J Med Res* 14 : 179-183

Gadgil RK and Shah SN. 1952. Human schistosomiasis in India. *J Med Sci* 6 : 760-763.

Gupta S. 2002. Clinical, biochemical and parasitological studies and prevalence of caprine schistosomiasis in and around Jabalpur. Ph.D. Thesis, Rani Durgavati University, Jabalpur.

Rathore BS 1998. An epidemiological study on buffalo morbidity and mortality based on four year observations on 18630 buffaloes maintained at 28 livestock farms in India. *Indian J Comp Microbiol.* 19 : 43-49.

Rollinson D and Southgate VR. 1987. The genus Schistosoma : a taxonomic appraisal. In Rollinson D, Simpson AJG (eds) "The biology of schistosomes from genes to latrines ". Academic Press, London.

Sumanth S,D'Souza PE and Jagannath MS. 2003. Immunodiagnosis of nasal and visceral schistosomiasis in cattle by Dot ELISA. *Indian Vet J* 80 : 495-498.

Vohra S. 2005. Development of immunological methods for diagnosis of schistosomiasis in small ruminants. Ph.D Thesis. Jawaharlal Nehru Agriculture University, Jabalpur.

WITH PRESENT KNOWLEDGE, CAN WE SUCCEED IN CONTROLLING SCHISTOSOMIASIS IN INDIA

M C AGRAWAL

Former National Fellow and Emeritus Scientist, College of Veterinary Science And A H, Jabalpur 482001

It is heartening to note that one part of this winter course is to teach molecular techniques about control of parasitic diseases. In the past also, there had been some advance courses and discussion about control of parasitic diseases. But whether we have been able to control any parasitic disease in reality ? Obviously, not. With changing time, it is imperative to know why we need to control parasitic diseases, which diseases should be given priorities, and how we should proceed to control these diseases. And this is what I am going to discuss ; thus my lecture will not be a mere academic exercise of citing references from some research papers. There is now time,we should discuss how we may try to control parasitic diseases and what should be our research priorities that may help in achieving these goals.

If we analyze disease control program of our country, it will be clear that the country has given top priority, logically, for controlling those infectious diseases that are affecting human population. Thus there are central government's programs for controlling tuberculosis, malaria, filariasis, guinea worm or naru,polio (small pox already eradicated) etc Interestingly, almost no attention is paid on zoonotic aspects of these diseases or medical faculty has not taken diseases where animals are playing a crucial role in their maintenance. For example, we have not taken sincere steps for controlling rabies. This reflects how wide apart are the two professions (medical and veterinary) which are supplementary to each other to combat zoonotic infections of the country.

When we come to animal diseases,the control programs have been launched only to limited diseases like Rinder-pest, FMD, anthrax with vaccination programs of these and other infectious animal diseases. These are the diseases,where veterinary hospitals,districts and states generate data and pass on to Government of India.

Has government of India ever launched any control program related to any parasitic disease of our domestic animal ? Here I mean control program like that of Rinder-pest or FMD. To my knowledge, there was or is none. No doubt there have been vaccination programs for controlling some parasitic diseases. The one was lung-worm vaccination for controlling lung-worm infection of sheep and goats,particularly in Jammu & Kashmir, Himachal Pradesh (now, vaccine production has been stopped by IVRI though infection is still existing). The other is *Theileria* vaccine produced by NDDB and supplied for controlling theileriosis among cross bred cattle. Again, this is not a infection which is affecting all our domestic animals. Thus we have attempted to control a infection (lung-worm) which is restricted to certain states or a infection which is affecting limited number of our domestic animals. And we have shied away in paying attention to the parasitic diseases which are affecting most of domestic animals and existing in all the states. Therefore, we have come to a crucial point where we need answer of alike questions - why we need to control parasitic diseases, which diseases are given priorities, and how we can control these parasitic diseases. My this short lecture will try to discuss some of these points with the hope that you, our young Parasitologists,will devote some time to find answers of these and related questions which are important for controlling parasitic diseases in the country. Answer of these questions are important as you will agree with me that all your research efforts will not fructify unless and until you attempt to control parasitic infections in our country, thereby reduce mortality and animal production losses.

WHERE IS THE DATA

If you go to any funding agency or even any government, the obvious question will be why we wish to control the particular parasitic disease. Naturally, you will put up the data about mortality and pathology of that parasitic disease. By the way, from where have you collected these data ? From some research paper or a review article or from a book ? And how they have calculated these data ?

I am sure that your data have not been collected from veterinary hospitals of the state or Animal Husbandry department of the country. And when these data have not been collected from the villages how they are representing correct scenario of the infection at village level ?

VETERINARY RECORDS

What I have said above, is all because of my experiences. While working on my book "Schistosomes and schistosomiasis in South Asia" (Agrawal, 2012), I tried to collect the data of nasal and hepatic schistosomiasis from veterinary hospitals and veterinary diagnostic laboratories (division level), state and central governments ; as, in my opinion, compilation of data from these sources is important since research journals fail to provide a detailed view of the infection ; these journals have not reported each and every outbreak or prevalence of the infection in each animal species from each geographical area.

To my dismay, I could not collect any data, related to schistosomiasis, from any of the above source- details are given in my book. Here, it will be suffice to mention :

- There is no direction from the government to keep record of parasitic diseases and details thereof. Neither any data are passed to state or central government.
- The veterinary hospital registers are maintained in such a way that you cannot extract any information from them. Even it is difficult to differentiate between new and repeat cases or species wise cases - cattle, dog and poultry.
- The veterinary diagnostic laboratories are vague in their reporting and counter checking is missing. While reporting coprological results (generally by direct wet smear method) positive cases are reported as "positive for helminth or fluke or roundworm eggs". Thus they have not differentiated the eggs where it was easy to differentiate.

It is clear that we need changes in above system and a fresh outlook is needed to the problem so that we will be able to collect the data related to parasitic diseases.

WHICH PARASITIC DISEASES TO CONTROL ON PRIORITY

This is an important question that if we decide to control parasitic diseases what should be our priorities ?

Answer of this question, at first hand, appears simple i.e. the parasitic disease which is causing greatest morbidity and mortality should be attended first. But how we will reach to this conclusion when we have not generated data in this regards. And when we are attempting to generate the data, are we using sensitive and specific methodologies ? This aspect is important else our results will not be accurate. While deciding importance of the disease, we have to examine its prevalence in each animal species with their age and sex, the seasons when disease is occurring, animal husbandry practices and geographical areas where infection is existing.

With the above, two more important aspects are also linked. One is the mortality occurring in the infected animals. It may happen, looking to Indian conditions, that the animal is not infected with single parasitic infection but multiple infections and in such cases we have to device methods for ascribing death etiology. The other aspect is related to pathology of the infection which is reflected to animal production losses. No doubt we will need new methodologies to decide the above facts and these may change as per advancement of our knowledge and environment. If we will not follow new techniques, there are all chances of making mistakes as has been done while reporting schistosomiasis in post-mortem cases (Rathore, 1998).

PRIORITY AMONG HELMINTHIC DISEASE

As my present discussion is related to schistosomiasis, I may raise a question to you - which is the most important helminthic disease of our domestic animals ? Some of you will answer - haemonchosis, or gastro-intestinal nematodiasis, others may answer fasciolosis, or amphistomosis; while some of you will say nasal schistosomiasis but a few or none will ascribe hepatic schistosomiasis as the most important helminthic problem of our domestic animals. May I ask the reasons for such conclusion ?

Those who are ascribing fasciolosis as most important helminthic disease will put forward their slaughter house experience where most of the livers yield adult *Fasciola gigantica*. Further, the fecal examination shows presence of *Fasciola* eggs in many cases (its different that we are not differentiating eggs of *Fasciola* from those of amphistomes). And those who are new to Indian parasitology will cite the reference of a foreign veterinary book where fasciolosis is considered as most important helminthic disease of domestic animals.

Those who are ascribing amphistomosis as most important will support their argument that every rumen in slaughter house possess one or other species of amphistomes and almost all fecal samples are positive for amphistome eggs.

Against these arguments,we do not have such visibility of hepatic schistosomiasis hence cannot be considered as most important helminthic disease in India. I have discussed this topic elsewhere (Agrawal, 1999, 2003) but will like to draw your attention to the need of employing correct techniques for reaching a correct decision. As time does not permit me to go into details of such discussion, I will mention only silent points emphasizing how important it is to use a correct technique and unbiased attitude for reaching to a correct conclusion. I am mentioning some facts for your consideration and further investigations :

- *Fasciola gigantica*, the only existing *Fasciola* species in India, spreads through *Lymnaea auricularia* or *Lymnaea accuminata* which is a strict fresh water aquatic snail (unlike to amphibian *Lymnaea truncatula* - intermediate host for *F.hepatica*) and survives only in perennial water sources hence animals will get infection only from perennial water sources.

- *F.hepatica* is more damaging to domestic animals than *F.gigantica*.
- Against one species of *Fasciola*, there are at-least five schistosome species (*Schistosoma indicum,S.spindale,S.incognitum, Orientobilharzia dattai* and *S.nasale*) which are affecting our domestic animals.
- Schistosome infections spread through *Indoplanorbis exustus* and *Lymnaea luteola* - both species are capable to survive not only in perennial water sources but also in temporary water bodies thereby giving higher geographical spread.
- A large number of amphistome species (40-42) are existing in the country (Dutt, 1980) some of which have been synonymized. However, only four genera have been incriminated in pathogenesis to domestic animals.
- In amphistomosis, it is only young animals which suffer and that too only from immature amphistomosis.
- Many fresh water snails are acting as intermediate hosts for amphistomes hence their geographical spread is as large as that of schistosomes.

Now you have to determine,using new methodologies, which helminthic disease is more important in our domestic animals. Here, you might have noted one more fact ; this importance will vary as per our husbandry practices. All these fluke infections are important only in grazing animals with limited effect in stall fed. Thus our priorities will vary as per animal species, husbandry practices and geographical areas. Therefore, you will identify those parasitic diseases to be given national importance and those with regional importance etc.

HOW TO CONTROL SCHISTOSOMIASIS

We have come to the last part of our lecture.And that is if we decide to control any parasitic disease in our domestic animal, how we can control it. I will take references from schistosomiasis which have been discussed in more details in my book (Agrawal, 2012). Indeed, this is the only helminth disease that has been given so much importance and where so many countries agreed to launch control program in collaboration with WHO (1985). Ironically, India has not participated in such program.

Interestingly, WHO have not touched zoonotic aspects of the infection while attempting control program of human schistosomiasis. My discussion will high light our short comings which are important to get rid of achieving any success in controlling any parasitic disease and also animal schistosomiasis in India or South Asia.

MULTIPLE APPROACH

No disease can be controlled following a single step approach, how so ever effective it may be. The infection has to be tackled from all aspects so that prevalence of it may be reduced to the levels where further transmission becomes difficult. Following multiple approaches were followed in schistosomiasis :

- Chemotherapy
- Health education
- Snail control through use of molluscicides
- Environmental modifications

You will realize that such program is not possible in India without a large team and funds which a department of Parasitology of an institute does not possess. No doubt, this WHO control program of schistosomiasis was able to reduce prevalence rate as well as pathogenesis, as the latter is directly proportion to the intensity of the infection; and the success was both due to political will and scientific team work. Here, it will be prudent to discuss scientific reasons for their success and how it will fit in our program of controlling animal schistosomiasis in India. One important difference has to be remembered that while following any control program,we have to monitor not one but more than one animal species.

CHEMOTHERAPY

The most important reason of the success, perhaps, was mass treatment of human population with a most effective schistosomicide (Praziquantel) whose single oral dose @ 40-60 mg/kg body weight was 80-95% effective in urinary and hepatic schistosomiasis. Since the weight of children varied between 20-50 kg, the dose of Praziquantel was cost effective, beside easy in administration.

When we compare this scenario with that of India, our schistosome species are different from those dealt by WHO hence there are all differences in schistosome bionomics. And praziquantel against any of these Indian schistosome species (*S.indicum,S.incognitum, S.spindale*) has not proved as effective as reported against *S.haematobium* or *S.mansoni.* In fact, our experimental work on pig- *S.incognitum* model has failed to reduce schistosomes to a significant level with presence of left over schistosome population as high as 40% (Shames *et al.*, 2000). A same scenario was observed when the drug was tried in other experimental models, with schistosome reduction ranging from 50-65% (left over fluke population between 50-35%), but using other Indian schistosome species. Ironically, no available flukicide (including anthiomaline) was found to lead to a significant reduction in schistosome population in any experimental model, tried in our laboratory or by other workers (Agrawal, 2012).

Therefore, it is not advisable to use Praziquantel or any other flukicide in any mass treatment program which has failed to reduce schistosome population to a significant level in our animals. This is because such use will lead to the high probability of development of resistance of left over schistosome population against praziquantel.

One more hindrance in using praziquantel for mass treatment in animal schistosomiasis is its cost for treating bovines against schistosomiasis. In one field trial against nasal schistosomiasis in Balaghat district of Madhya Pradesh, 60 tablets of Prazi Plus®, which works out approximately to 20 mg/kg body wt was effective in eliminating completely clinical symptoms of snoring disease from 14 out of 18 cattle. However, cost of one treatment comes to Rs 1500/per animal as market cost of two tablets is Rs 50/ (Agrawal, 2012).

DIAGNOSTIC METHOD

While implementing any control program, it is important to monitor the rate of infection and its intensity in the given population prior and after the treatment. This monitoring is possible by using a simple, rapid, economic but sensitive parasitological diagnostic technique that may detect, light,moderate and heavy infections. In WHO case, Kato-Katz technique was applied that fulfilled all the above criteria and was able to detect even light infections (24-96 epg) with presence of 1-4 eggs/slide, made of 41.7 mg of feces (WHO, 1985). In contrast, a very low epg, ranging between 2-4 in ruminants and 10-50 in pigs has been reported by almost all the Indian workers (Agrawal, 2012) ; for this reason almost all egg detection methods have poor efficacy,cannot be employed in control program, hence warranting further work on this issue.

SNAIL CONTROL

The snail control,in WHO program, was undertaken by using Niclosamide which has proved highly effective in killing the intermediate hosts i.e. *Bulinus* and *Biomphalaria*. Ironically, Niclosamide is not available in Indian markets and has to import from other country. We were able to procure Niclosamide and undertook both experimental and field trials on fresh water snails, observing its high efficacy in killing *Lymnaea luteola, Indoplanorbis exustus* and *Gyraulus* but without affecting other water fauna and flora (Agrawal *et al.*, 2005, Agrawal, 2012).

However, ecological conditions of fresh water snails in India,differ from those of other endemic countries. In India, these are the ponds, tanks and other water sources, both perennial and temporary, which are inhabited by these fresh water snails, where as in endemic countries these were the rivers where snails are surviving. Thus, we have more wide spread population of the snails, making our efforts more difficult while attempting snail control.

BIOLOGICAL CONTROL

As use of chemical molluscicides may cause environmental pollution, efforts are being made to find other alternatives for snail control. Work is being carried out to find Plant molluscicides which are more eco-friendly. Likewise, search is being made to find natural enemies of the snails that may be used as biological control agents for reducing snail population. However, it must be remembered that the biological enemy or competitor should not be introduced from outside the country/area as they may prove harmful afterwards as experienced by WHO (1985).

One natural enemy of young fresh water snails has been identified by us (Agrawal, 2012) in the form of nymphets of dragonfly which co-habitat with the snails under same ecological conditions ; these nymphets are carnivorous in nature, with life span of 1-3 years; devouring young snails hence there is no need of re-introduction every year. There is the need to develop the techniques for mass production of these nymphets which also feed on mosquito's larvae (Agrawal, 2012). Here is the importance and need of your molecular techniques which may pave the way of producing eggs and nymphets in large number, for their use in controlling snails or mosquito larvae.

HYPER-PARASITISM

As stated above, Praziquantel and anthiomaline are not able to kill a significant number of Indian schistosomes. There are also chances of developing resistance in left over schistosome population after wide use of these less effective drugs. Therefore, there is the need to think alternatives against chemotherapy for parasitic diseases.

One promising area of research might be searching hyper-parasites of schistosomes or of any other parasite which is causing disease in man and his animals. As you are aware these hyper-parasites are those organisms which parasitize on the parasite itself e.g. schistosomes. This hyper-parasite may be a bacteria or a virus or even a fungus. We have to search those hyper parasites which are lethal to these parasites but should be host friendly thereby eliminating our animals or human beings from any harm.

With this lecture, I tried to discuss how ill equipped we are for launching a state or national level control program against any parasitic disease of our domestic animals. There is the need to re-think on our research programs and to develop methodologies which may help directly or indirectly in controlling parasitic diseases of the nation. But it will be a bad idea not to take any control measure and waiting for developing proper techniques or drugs prior launching any control program. Notwithstanding these facts, management of animal excreta may greatly help in controlling parasitic diseases. If we start collecting all cattle dung to fill our gobar gas plants or to convert it into vermin-compose, this will not only control parasitic diseases but will also improve rural economy.

At last, I once again emphasize the need of a National Institute of Parasitology in country for paying proper attention on the parasitic diseases,which are causing not only mortality but also production losses thereby affecting adversely national economy. I am avoiding repetition what I have already said during my lecture of Dr SC Parija oration gold medal award in 2012 at Indore and which has been published in Para Sight 2 Volume 2,issue 1 of IAAVP and is also posted on my blog www.indianschistosomiasis.blogspot.com

SUMMARY

Launching of control program for any animal disease which is causing mortality and production losses is important for any country. Ironically, the government of India has launched control programs for viral and bacterial diseases but there is no control program,ever launched by any central or state government against any parasitic disease of our domestic animals. Neither, there is any system of collecting data,related to animal parasitic diseases, from our veterinary hospitals and veterinary diagnostic laboratories. Therefore, it becomes difficult to answer- why we need to control parasitic diseases, which diseases are given priorities, and how we can control these parasitic diseases ?

When we come to the question of how we can control animal schistosomiasis in India, we observed that our diagnostic technique is not sensitive due to low egg production of Indian schistosomes and requires further research. Likewise, the most effective Praziquantel drug is not able to kill significant number of Indian schistosomes with 35-50% left over blood-flukes. To circumvent this situation, it will be prudent to find out hyper-parasites of schistosomes and other parasites which will selectively kill these without harming their hosts. However, it will be futile to wait for developing proper techniques. Instead, animal excreta may be collected at mass scale and may be used in gobar gas plants or for preparing vermin-compose which will not only reduce prevalence of parasitic diseases but will also improve rural economy. The author has again emphasized opening a national institute of parasitology so that a holistic view may be developed towards parasitic diseases,existing in the country.

REFERENCES

Agrawal, M.C. 1999. Schistosomosis : an underestimated problem in animals in South Asia. *World Animal Review* 92: 55-57.

Agrawal, M.C. 2003. Epidemiology of fluke infections. In (ed Sood ML) Helminthology in India. International Book Distributors, Deharadun. Page 511-542.

Agrawal MC, Singh KP, George J and Gupta S 2005. Niclosamide trials on *Indoplanorbis existus* and *Lymnaea luteola* under different conditions. *J Parasit Dis* 29: 53-58.

Agrawal MC. 2012. Schistosomes and schistosomiasis in South Asia. Springer India Pvt Ltd, New Delhi.

Dutt SC 1980. Paramphistomes and paramphistomiasis of domestic animals of India. Punjab Agriculture University, Ludhiana. Page 162.

Rathore BS 1998. An epidemiological study on buffalo morbidity and mortality based on four year observations on 18630 buffaloes maintained at 28 livestock farms in India. *Indian J Comp Microbiol.* 19: 43-49.

Shames N,Agrawal MC and Rao KNP 2000. Chemotherapeutic efficacy of praziquantel and closantel in experimental porcine schistosomiasis. *Indian J Anim Sci* 70: 797-800.

WHO 1985. The control of schistosomiasis. Technical report series 728. World Health Organization, Geneva Page 113.

SOME INTER DISCIPLINARY AREAS

With advancement of research on parasitic diseases, scientists have realized that it is not possible to solve a chronic or new problem by an individual scientist but it may require a team of scientists, gathered from different disciplines of sciences. These researches have given new directions to the science with development of different branches of parasitology for finding solutions of some old or new problems. At times, this joining of the two disciplines are from quite unrelated subjects like history and parasitology when the team is trying to study impact of a parasitic disease on retrieval or defeat of a country or to know whether Malaria was the reason for death of Alexander the great or geography and parasitology to study how geographical changes were responsible for evolution of a particular parasite. We have mentioned below some important new emerging research areas where Parasitologists have become interested to investigate the problem in association with scientists of other discipline.

1. **Biotechnology and Parasitology** : Biotechnology is a new branch of biological sciences which manipulates genes of the organism and carry over different works -hybridoma technology has been very popular. Biotechnology has immensely been used in modern parasitology to understand various problems. PCR is the new technique for identifying even nano quantity of a parasite in a host species, thereby giving parasitological diagnosis a new dimension and helping in identifying parasitic infections in different host species including various intermediate hosts. It is extensively being used in developing diagnostic molecules, vaccine molecules and unraveling genome of the parasites.

2. **Genetics and Parasitology** : Perhaps, the first evidence of its importance came when Hsu and Hsu proved presence of strains of *Schistosoma japonicum* (50) which have different susceptibilities to their final hosts. Further studies showed that a parasite species may not be a homogenous mass but may have different strains, hybrids, subspecies of parasites which have different susceptibilities to intermediate and final hosts. The genetic make up of the hosts also affect their susceptibilities to a parasitic species and this led to work on developing sheep, cattle breeds, resistant to *Haemonchus*, trypanosomes etc. Genetics has made great roads in parasitology in every sphere and there are many studies which require genetic knowledge. Phylogenetics has helped to understand parasitic evolution. Studies have also been made to understand whole genome of important parasites and presently whole genome is known for

many important parasites like *Schistosoma mansoni, S.haematobium, Necator, Plasmodium* etc (49).

3. **Immunology and Parasitology** : The first hurdle whenever a parasite enters in a host is fighting with innate resistance of the host and later acquired resistance and both are determined by host immune system which differs as per host species. How the parasite is circumventing this immune system or how at times host is developing autoimmunity or immuno-pathology in its attempts to get rid of the parasite is fascinating work being carried out by immunologists in collaboration with Parasitologists. How immunological systems of different animal species are responding to different parasites is being worked out. Interestingly, even same host species is found susceptible to one species of same genus and resistant to other i.e. cattle are susceptible to *S.indicum* but not for *S.incognitum (2)*. Immunology is also helping in developing new immunodiagnostic methods, kits, vaccines against important parasites like malaria, schistosomes, coccidia, *Theileria* and filaria (infact, this field is more difficult and complicated than developing vaccines against bacterial or viral infections).

4. **Biochemistry and Parasitology** : A great interest has developed to understand basics of many biological phenomena occurring in parasites and its hosts and this has resulted in development of biochemical parasitology. It was able to explain role of chemicals in trigger mechanisms, metabolic pathways in different parasitic stages, biochemical structure of parasites, biochemical targets, bio-physiology of parasites, biochemical changes occurring in infected hosts and in resistant strains. More and more work is being carried out by the two groups of the scientists to understand biochemical basis of important phenomena.

5. **Nutrition and Parasitology** : Research has shown that tapeworm infection of *Diphylobothrium latum* in man absorbs ten to fifty times more Vitamin B^{12} causing vitamin deficiency anemia. Hook worm infections (*Ancylostomum, Nicator* sp) are responsible for causing severe anemia that may be fatal in children and pregnant women. Our research has revealed that schistosomiasis is causing hypoglycemia in infected animals (2). All these and other experiments have shown that the parasites are causing nutritional deficiency (minerals, vitamins, hormones etc) in the host hence it is important to supplement the chemotherapy with deficient nutrient as well. Inversely,protein/carbohydrate/minerals/vitamins etc deficiency is adversely affecting immune status of host species for a

given parasite leading to more damage to the hosts. All such studies are conducted by the team of scientists from the two disciplines. Though clinical nutrition is a well developed discipline in medical sciences, there is the need to develop it in veterinary science as well.

6. **Pharmacology and Parasitology** : For curing a suffering patient, a drug or chemotherapeutic agent, is essential without which it is rather impossible to cure the patient. Here comes the role of pharmacologist who helps in developing new drugs and understanding pharmacodynamics or how a drug behaves in a host species. There are only limited drugs for parasitic infection and development of drug resistance by the parasite is an emerging problem. A combined effort is continued in screening medicinal plants, chemical molecules against parasites, snails, insects.

7. **Parasitism and Hyper-parasitism** : This is for searching hyper parasites in the form of bacteria, virus, fungi which parasitize on important parasites and to identify those hyper-parasites which are able to kill the parasite without causing any pathogenic effect to the host. There have been now close links with microbiology and parasitology to find such hyper-parasites. This topic has gained importance in light of limited availability of drugs and reluctance of our pharmaceutical companies to search new drug molecules against parasitic infections due to limited market value. It is important to mention that *Bacillus thuringiensis* (*Bt*) has been developed and marketed commercially which is lethal to mosquito larvae and thereby this hyper-parasite is helping in controlling mosquito population (51).

8. **Synthetic Biology** : This is a new branch of science which involves creation of biological systems intended for specific purpose. According to an article "Scientists create life with alien DNA" by Andrew Pollack and appeared in the Economic Times, Indore, MP, dated 9th May, 2014, page 11 (Research Paper Published in "Nature") the scientists of Scripps Research Institute, USA have been able to insert two chemically synthesized nucleotides, which they called X and Y, into the common bacterium *E.coli* already having natural four nucleotides; the bacteria were able to reproduce normally, replicating X and Y nucleotides also along with the natural nucleotides. This has been first time when a living cell manage an alien genetic alphabet and will have far reaching effects in our biological sciences as well ethical, legal and regulatory implications. It will not be out of contest to mention that despite the great diversity of life on earth, all species from human to virus use the same genetic

code made of these four natural nucleotides. As all the products, produced by a cell depends on its genetic makeup, scientists hope to create new organisms having enormous powers including producing new type of antibiotics, biological products that natural cells cannot produce. However, a danger also develops of creating some new type of bacteria and viruses that may cause havoc to the life existing on the earth.

9. **Geology and Parasitology** : Research has shown how geological topography of an area is important for survival of the parasite-including formation of geological plates and river and pasture formations, etc A new branch of Paleoparasitology is developing which is trailing evolution of a parasite and linking it with human evolution as well as geographical developments (52). A good example is about evolution of schistosomes and why some schistosomes are infecting human beings and others' not (Agrawal, 2012). Geological studies and development of rivers have also been under consideration while studying how schistosomes and other parasites spread to different continents. In Zoo-geology, scientists are relating existence of the parasite in a particular geological area because of specific environmental characters.

10. **Distance Imagination and Parasitology** : Presence of intermediate hosts like insects, snails and wild animals have close link with ecological conditions like presence of water bodies, rivers, trees, rocks ; distance imagination work is helpful to take photographs and other details of geographical areas through satellite which are otherwise not reachable. By studying these photographs and other details, it is possible to know which snail species, insects, wild animals are existing in these areas with probable diseases transmitting under given conditions. This study is a warning for chances of spreading new pathogens whenever man or his domestic animal tries to venture in such areas for different reasons.

11. **Mathematics and Parasitology** : Beside the role of statistics in understanding significance of the findings, mathematics is playing a crucial role in deciding thresh hold levels for survival and transmission of a parasite, determining parasitic population in a given population and developing new mathematical formulae in understanding epidemiology of parasitic diseases.

12. **Economics and Parasitology** : The parasitic diseases affect human economics in two ways. The first is suffering of human beings, loss of their life and working days there by affecting economics of the

family. For example, *Onchocerca volvulus* is a filariid that infects humans of African countries and as the infection advances, microfilariae of the filariid accumulate in large concentration in eyes leading to blindness and migration of young population to other places. This has disturbed socio-economic behavior of these communities. Likewise, presence of trypanosomes in these countries has caused sleeping sickness in human beings,thereby affecting socio-economic framework. The Second is loss of animal production and quality of animal products due to parasitic infections. Trypanosomiasis in animals have hindered animal husbandry development in some African countries though the places are most fertile regions. Therefore, greater studies are being conducted to understand how parasitic diseases are affecting economics of a nation.

13. **History and Parasitology** : Now there are critical studies to know if parasitic diseases like malaria, filariasis, schistosomiasis have ever played any role in defeat of an army or its retrieval ; if any historical heroes ever died of important parasitic diseases.Alexander the great came near the Indus river and returned back; but within a short period of his return, he expired in the young age of 28. There are various studies going on to know reasons for his death ; a study is going on to know if he and his army suffered from malaria which they might have acquired in India. Schistosomiasis is an old infection in Egypt as its eggs have been demonstrated in the mummies ; now work is going on to know whether the infection shortened lives of some Pharaos (kings) and how much this infection was prevalent in their kingdom. Historical studies are being undertaken to know what role was of a parasitic disease in wiping out human or animal population of a given town or locality.

14. **Forestry and Parasitology** : *Trichinella spiralis* was circulating among wild boar and bears in the forest and when man invaded forest for hunting and eating their flesh, he suffered from trichinellosis. Schistosomiasis, paragonimiasis, hydatidosis,trypanosomiasis are other examples of parasitic diseases which were cycling among forest animals and man got these infections by encroaching these territories. There appears little work to study forest ecology, animal population dynamics which were regulating these and other diseases under natural conditions.Understanding these conditions have become of concern since man is constantly interfering in the forests either for deforestation or construction of dams, roads etc and acquiring these and other infections.

A sad commentary on parasitic diseases is that it is mainly a problem of rural population of developing countries or tropical countries who are struggling to enhance life expediency of their population with increase in their economy which indirectly depends on well being of domestic animals. There is great scope to contribute for up gradation of human life by solving problems related to parasitic diseases. No one knows what new comes out by the research efforts of our talented young generation hence those who are interested to join research as a career may join Parasitology as every field of research is said to have "Sky is the limit."

10

Search of A Guru

It is always advantageous if there is a senior scientist,a mentor or a Guru (I have not used the term teacher as all may be teachers but may not be Guru as per Indian mythology) who can recognize your talents and may believe that you may achieve some unusual goals, if properly guided or encouraged. This is important for carrying an important task to its fruitful results. Here what we are referring a Guru is not in the sense as is presently used for searching a god father. There is a vast difference between the two. The perception about god-father is that he may not possess the required knowledge but is perfect in manipulating situations in his favor so that work is done by all means. In scientific world, this is mostly applied to the persons who have contact with higher ups helping in getting promotions or foreign trips. These god fathers are good only to provide some material gifts but are not best in their scientific achievements hence cannot replace a Guru or a good scientist.

The history tells us that finding a correct teacher or a mentor has many times changed the life of an ordinary person. This appears true everywhere in our life, not alone in scientific world. If we look our ancient history, it was Chankya (53) or Vishnugupta who metamorphed a simple child in an almighty king Chandra Gupta and was instrumental in defeating mighty king Nand of Patliputra, Bihar. In the literature, there are many instances how the poem or story of a new comer was corrected by a senior literary person or how new comers have been encouraged to publish their poem, stories in famous literary magazines by the renowned editors of these magazines.

In the research fields also, there is a need of a Guru who can recognize talents within you and guide you properly so that you may excel in your field. Like a good disciple, there are certain characters which a teacher should possess to be a torch light for you. First is his deep knowledge of the subject where you wish to excel. It does not mean a bookish knowledge but a practical one which is reflected in his excellent research publications and his awards which he won in scientific career. The other important character is that the teacher is willing to impart his knowledge to you without any restrictions i.e. he has considered you a disciple who will continue his mission even afterwards. A great love and affection between teacher and disciple is necessary to carry over big achievements. And being a knowledgeable person he should be pleased to answer your queries ; even if a time comes, when your work reveals that his some findings are erroneous,he should be glad that his student has attained such a height and intelligence to contradict the findings of his predecessors. If any one of your teachers possess such qualities, you should remain in touch with him and be guided by him.

There is some criticism of our students and Indian culture of respecting their teachers and not questioning their scientific findings. I think this is a short sighted conclusion without understanding Indian culture. The historical facts do not support such conclusion. It's correct that an Indian student gives full regards to his teacher or Guru who has taught him the subject but these teachers have never hindered their students' thought process- rather they always encouraged to think something new. When a student joins research field, he is encouraged to think new and there is no restriction about his arguments or criticizing findings of his teachers or any other scientist. And this is the reason that Indian philosophy developed with so many varied ideas and creation of many schools in Indian culture. These were the erstwhile students of some prominent Guru or Rishi of those time who promoted new ideas and brought new conclusions enriching Indian culture with many school of thoughts.

If we go into the history of science, there are many examples where Guru was responsible to highlight the talents of his disciple or it was the peer whose efforts brought the achievements of their juniors/students in the light of scientific world. And this was the greatness of the mentor that due credit was given to real person. The best example,we may cite, is that of Srinivasa Ramanujan (21), the great mathematician of twentieth century whose birth day -22nd December –is celebrated as National Mathematical day, in India. Ramanujan was born on 22nd Dec 1887 in Erode,Tamil Nadu in a orthodox Tamil Brahmin family. He was extra-ordinary in mathematics but poor in other subjects hence could not maintain his scholarship in Kumbakonam government college. He joined another college to pursue his mathematical interests and to sustain himself he worked as a clerk in accountant general office of Madras port trust office. Being a genius,he was doing many mathematical logarithms but without any answer. Therefore, in 1912-13, Ramanujan dispatched samples of his theorems to three academicians of University of Cambridge and one of them was Prof G H Hardy. Credit must be given to Prof Hardy who could recognize brilliance of this work and persuaded Ramanujan to join him in Cambridge University which he did.This resulted in publication of extra-ordinary contributions to mathematical analysis - number theorems, infinite series and continued factors. In strict sense, we may not call this relationship as that of teacher and taught but great credit goes to GH Hardy who solved the puzzles of Ramanujan and took all the pain of publishing them but giving due credit to Ramanujan hence earning fame for both.

The other example that may be cited is of Satyendra nath Bose (54) and his teacher Albert Einstein under whom Bose took his Ph.D. degree. After coming to India, Sateyndra Bose suggested the theory of sub-atom and his teacher took pain to translate his research paper in German and to publish it in a German research journal. It may not be out of place to mention that the recent research has confirmed Bose's hypothesis of existence of sub-atomic particles or what has been named as God particles (or Boson particles).

More close to Biological sciences is the example of Sir Ronald Ross and his mentor Prof Patrick Manson who is regarded as father of tropical diseases (23). Ross, a medical officer in Indian army, took the leave to pursue his Public Health degree from London. During London stay, Manson showed him Laveran's bodies in human blood and convinced Ross that these are malaria parasites (as earlier claimed by the French scientist Alphonse Laveran) which is a protozoa and appears a causative agent for malaria. As Manson was working on filariasis at that time and suspecting role of mosquito in spreading the disease,so he suggested that malaria might also be carried away by the mosquitoes. However, at that time, it was not believed that insect bite may transmit any infection and Manson was suspecting transmission by drinking the infected water. So here also, Manson suggested that malaria might be carried through drinking infected water.

Ross came back to India, conducted experiments of breeding mosquitoes, feeding them on human patients,mixed them in water and fed to humans but this hypothesis failed. Ross wrote to Manson "the belief is growing on me that the disease is communicated by the bite of mosquitoes...." And after some more failures,Ross was able to demonstrate that mosquito bite is responsible for transmission of malaria and it was Prof Patrick Manson who informed the Royal Society about this discovery of Sir Ronald Ross. Here, you will appreciate that Ross followed what Manson has suggested but it was not a blind following and when he failed to transmit malaria by drinking infected water, he was compelled to think to test if biting of mosquito will transmit the disease and received the success.

It may not be out of context to mention that I too was initially not interested to pursue my studies in Parasitology and it was my Guru late Professor SC Dutt (55) who taught me the subject with great interest and was instrumental in my learning basics of Helminthology.Therefore, it is the reason to believe why our ancient sages have given much emphasis and importance to the Guru in the life of a person who can change life of his disciple.

11

Suggestions by the Peers

If you have prepared yourself to make research as your career, it is most advantageous to read research papers/reviews/chapters/books published by the eminent scientists. It is not necessary to confine reading to your subject but a voracious reading will enable you to develop an idea about general problems related to science and our society; how scientists managed to solve them and with development of new knowledge, how you may tackle fresh or existing problems. A deep analysis may be done of those works which are so accurate that even after lapse of so many years, they still hold true. In our opinion, it is better to launch websites/blogs highlighting important works of eminent Parasitologists for the benefit of our research scholars; a beginning, in India in Parasitology, has been made in this direction with launch of a website on first time Rafi Ahmed Kidwai award winner (in Animal Sciences) late Dr SC Dutt (www.drscdutt.com). This website (now discontinued due to hacking problem) and two blogs (www.indianschistosomiasis.blogspot.com www.indianparasitologists.blogspot.com) contain scientific matters, a write up,critically analyzing why,even after lapse of fifty years of the research, the findings of Dr SC Dutt have still not been disputed by his successive scientists.Beside these, there are the websites of the three Parasitological societies of India providing many details in relation to India; likewise, all international parasitological societies and other scientific organizations are having their own websites which are worth reading by our Parasitologists.

It is a general tendency in scientific community to accept the findings of their peers without making critical analysis. However, review of research papers has confirmed that all the findings and suggestions, even of the peers, have not always proved correct. Nevertheless, accepting these results on their face value deter the future scientists to believe otherwise. Moreover, these reports have inbuilt risk of giving wrong direction of future research thereby taking long time for correcting the peer's hypothesis. I may explain this by an example which I am well aware.

As mentioned earlier, Dr H D Srivastava, Head,Division of Parasitology, Indian Veterinary Research Institute, Izatnagar and his associate Dr S C Dutt started experimental work on Indian schistosomes in 1950 and onwards. These scientists became authority on Indian schistosomes due to their tremendous original contribution on schistosomiasis and still regarded doyens of the subject. They have guided many students for their doctorate program and one student Dr P K Sinha of West Bengal Veterinary College, Calcutta did his Ph.D. degree in 1954 under Dr Srivastava. As not much was known at that time on Indian schistosomes, Dr Sinha, in his doctorate program,investigated life cycle of *Schistosoma incognitum* including susceptibility of different hosts to the blood fluke. Many of these findings still hold true and referred by contemporary scientific community. For checking susceptibility of different animals, including albino mouse, they infected these animals with *S.incognitum* cercariae (300 to 1000 cercariae per mouse), checked their feces for excretion of fluke eggs and confirmed presence of adult flukes and its eggs in host's body,after sacrificing them within stipulated period (56). They recovered adult male and female schistosomes from albino mouse with presence of schistosome eggs in liver and intestine but failed to detect them in mouse feces. As fluke eggs, fully developed with presence of miracidia, were present in liver and intestine but could not be detected in the feces, Sinha and Srivastava (56) suggested " in white mouse, though the worms attain maturity, their size was comparatively much smaller and their eggs could not break through the wall of its intestine to be voided with the feces". Dr S S Ahluwalia was also working for his Ph.D. program (1968) at IVRI, under Dr S C Dutt and used *S.incognitum*-mouse model as part of his research program. He too could not detect eggs of *S.incognitum* in the feces of mouse, hence supported the view of Sinha and Srivastava.

It is important to mention that finding fluke eggs in feces is an important tool being used for assessing drug efficacy or immunity in the host against the infection. This phenomenon of excretion of eggs provides many other related parameters for judgment i.e. pre-patent period, patent period, fecal egg concentration which could not be considered in this *S.incognitum*-mouse model.

In early seventies, many Helminthologists, in other countries,were working to check development of immunity against schistosomes (other than Indian species) and were using mouse model being an easy experimental model. In India, Dr S C Dutt, at Jabalpur Veterinary College attempted to replicate the experiments to check if Indian schistosomes also develop acquired immunity and selected *S.incognitum*-mouse model for this work. His many students undertook their theses problems using this model for checking development of immunity by pre-exposing the host with schistosome cercariae. As it was believed that *S.incognitum* eggs do not pass in mouse's feces, all the related criteria could not be considered; the only one left was to analyze fluke load in two groups of mice (immunized and challenged, and challenged alone) while a few scholars also compared size of the schistosome in the two groups. I also worked on heterologous immunity in 1975 using mouse model but did not consider fecal egg load due to above reason though I compensated this criterion by taking additional parameters like number of schistosome eggs and number and size of granuloma in liver of mice of the two groups. Where ever this *S.incognitum*-mouse model was used in any experiment, including chemotherapeutic trials, fecal egg excretion could not be undertaken. Thus one can appreciate how one observation by an authority has changed direction of our research work and this status remained so continuously for twenty years.

This situation changed in 1985,when we undertook work on diagnosis of schistosomiasis in final hosts, under an ICMR research scheme. This time, we infected albino mouse, 12 in number, each with 500-600 *S.incognitum* cercariae (57). As this blood fluke attains maturity in albino mouse in about four week time, we started examining its feces,on group basis, from 30 days post infection. There are various methods of fecal examination for detecting schistosome infection; the most simple but less effective is direct smear examination while acid -ether and hatching methods are among more sensitive diagnostic methods. Therefore, we followed latter two methods in our endeavor. Coupled with this, we started sacrificing the infected mice from 36 days onwards to record number of fluke pairs as well as concentration of viable schistosome eggs in liver and intestine. Though, eggs were present in these two organs, its concentration in intestine was low on 36 day post infection that increased,afterwards, with passing of time. On 49th day of infection, our acid-ether method could detect *S.incognitum* eggs in the pooled fecal material hence we examined feces of each mouse on same day and onwards. All mice,except one, turned positive for the eggs on 49th day itself;the lone negative case also turned positive on 54th day. The eggs remained present in feces till last day of observation I. e. 71-73, when last mouse was sacrificed. The egg per gram of feces ranged between 300 to 1000 with a Mean of 400 epg (57).

If we analyze above experiments,we selected two most sensitive diagnostic methods for above work. The hatching method is a specific diagnostic method which is simple and most of times proved more sensitive to any egg detection method, including acid-ether in domestic animals. Thus it was logical to presume to find miracidia while applying hatching method in above case. But this did not occur in mouse thereby showing it not an efficient diagnostic method in this model. However, it will be wrong to think absence of hatching of eggs in mouse feces; the phenomenon is simply highlighting poor efficacy of the method and a warning on the perception of obtaining identical results in different host species.

What may be the reasons of failure of detecting *S.incognitum* eggs in mouse feces by previous workers ?. Prior to this analysis, it is pointed out that pig is a confirmed natural host for *S.incognitum* where the fluke attains maturity within four weeks of the infection. The pig starts excreting the eggs in the feces on 30th day of infection or so and is detected by hatching method and also by direct smear method. Since mice were also developing mature schistosome infection within four weeks time, a simulating situation of excretion of fluke eggs in its feces might have presumed with applying similar diagnostic methods. As pig feces turns positive between 30 to 36th day of infection, they might have examined mouse feces continuously for a week, starting from 30th day. When these scientists failed to detect any schistosome eggs or miracidia in mouse feces, they tend to believe that mouse is not excreting schistosome eggs in its feces. The success of our experiment was by extending the observation period which in turn was relied by observing lower egg concentration in the intestine,even after 36 day of the infection. Additionally, use of acid-ether instead of direct smear enhanced efficacy of the diagnostic test enabling to detect even low concentration of fluke eggs.

In the beginning of twentieth century, there was intense interest in schistosomiasis research in India. The scientists (2) failed to recover female *Schistosoma spindale* from the guinea pigs resulting in suggesting by Fairley in 1930 that there are certain "host factors" that are preventing development of female blood flukes in the guinea pigs and identification of these host factors may help in controlling the infection. This suggestion by the peers was carried away by subsequent researchers. After about 20 years of this hypothesis, this was the work of Professor SC Dutt (18), done meticulously on the guinea pigs, that provided contradictory evidence of developing female *S.spindale* with laying of eggs by the fluke in the liver of the guinea pigs.

Another example of wrong observations by the peer is that of Montgomery (58) who discovered three schistosome species i.e. *Schistosoma indicum, S.spindale* and *S.bomfordi* in 1906 and described morphology of each blood fluke but considering all of them being tuberculated- having tubercles on their cuticle. This possession of tubercles by *S.spindale* created great problems in identifying etiological blood fluke of nasal schistosomiasis (*S.nasale*) which was also tuberculated but differed in its location (2). Again a suggestion was put forward to exist two types of *S.spindale*- one tuberculated and another non-tuberculated.After about fifty years of discovery of *S.spindale* by Montgomery, Dutt (18) conducted experimental work on *S.spindale* and concluded them as atuberculated or with smooth cuticle. The electron microscopic work and work of other workers (2) conclusively proved *S.spindale* being a non-tuberculated blood fluke. It may be interesting to note that we do not have details of *S.bomfordi* or *Orientobilharzia bomfordi* (reported from cattle) as no other worker has ever encountered this parasite in any host species although a somewhat similar blood fluke namely *Orientobilharzia dattai* has been well described from cattle. The main difference between the two blood flukes is while former is tuberculated, the latter is atuberculated. Is it again an erroneous observation making *O.bomfordi* a tuberculated fluke and obscuring its identity?

We have mentioned in chapter 6 how science has grown. There also you have observed the hypothesis of spontaneous origin of life put forward by Hippocrates and later Aristotle who believed that life generates from nothing or spontaneously. Again germ theory for communicable diseases was proved only in 1880 by Luis Pasteur and earlier scientists believed that this is bad air (Miasma theory) which is responsible for spread of diseases.

Perhaps, this is the way of progress of science where a stalwart puts forward some suggestion that is carried away for centuries together and suddenly there enters a new genius who questions the existing hypothesis and put forward a new idea about that subject. This progress of science re-confirms our philosophy that never accept any idea as gospel truth and you are free to test any hypothesis which does not appear true to you.

12

Think for A Change

One prime fact in research is not to take the matter for granted and do not believe in *status quo*. This is most critical mindset for creating something new. Even change may be made in existing circumstances with minimum requirements. Here, I may cite my experience while starting research in 1976 for my doctorate program on immunity against schistosomiasis. The experimental model was albino mouse where schistosomes, by their nature, develop into blood vascular system of the host hence require a special technique called "Perfusion technique" for retrieval of mature and immature schistosomes. This 'perfusion technique' requires use of a "Peristaltic pump" and was followed in some western countries (2). However, No body in India followed this perfusion technique hence recovered schistosomes from mouse or from any other host species by teasing the blood vessels or/and cutting and soaking liver,spleen, kidney, lung pieces in normal saline. After about 4-6 hr, the tissue pieces were removed, excess blood-saline was decanted and remaining whole material was searched, in small aliquots, under a stereoscopic microscope for presence and picking of schistosomes. This "cutting and soaking" is good for studying life-cycle of any schistosome species but not appropriate for immunological or therapeutic experiments where accurate recovery of schistosomes of the two groups is a prerequisite for reaching to any conclusion. Moreover, as only saline is used, without any anti-coagulant, some schistosomes are entangled in the blood -clots making the fluke recovery more questionable. Even then, all the previous Indian scholars, prior my work, used this 'cut and soak' technique in all the experiments related to the schistosomes (2).

I was not satisfied with present scenario and started searching available literature in our library. This search revealed use of above perfusion technique employing a Peristaltic pump in recovering schistosomes from experimental models. Rather concentrating on the equipment, I tried to understand the principle of the technique which was to force anti-coagulant (sodium citrate) mixed normal saline at a rhythmic manner in the posterior vena cava and to recover the blood mixed citrated saline or perfusate from severed hepatic portal vein. This method of recovering the schistosomes from the host was quick and more accurate than our 'cut and soak' method. However, there was no Peristaltic pump in whole of the Veterinary College ; neither there was any possibility to purchase this imported equipment due to paucity of funds and other hurdles. My search revealed presence of an "Automatic Pippeting Machine" in Microbiology department which was donated to the department under USAID. Use of the machine was to dispense measured quantity of bacterial culture media (molten agar etc) in Petri-dishes or culture tubes at constant speed and quantity of culture media may be adjusted using appropriate mechanism. As the machine was dispensing measured quantity of the fluid with a force at constant speed, I considered it worth trying in perfusion technique for recovering schistosomes. Since it was difficult to use posterior vena cava in albino mouse due to its small size, we used right ventricle of heart to initiate perfusion. As we followed the principle of the 'Perfusion technique' correctly, our modification was successful (59) and followed by subsequent workers. This led to recovery of schistosomes more accurately and in shorter time without following cumbersome processes.

However, this automatic Pippeting machine could not serve the purpose when we tried it for perfusing large animals warranting need of a change for perfusing large animals like piglets, lambs or calves ;here we required to infuse larger quantity of citrated saline with more force and constant speed and this Pippeting machine had limitations. Moreover, there is breakage of glass pipettes, used for dispensing citrated saline, in spite of the precautions. Therefore, I was in search of any equipment which may disperse larger quantity of saline at rhythmic speed. This led my attention to vertical water pumps, commonly used in desert water coolers in north India during summer. But I was skeptic if it may serve our purpose. The first doubt was because its ejecting water constantly while automatic Pippeting machine ejects at regular intervals-with gaps ; there is a great force in dispensing the fluid by the water pump and whether host's blood vascular system can sustain such force or it will rupture the blood vessels ; if there are chances of damage of water pump by using sodium citrate and sodium chloride instead of water. Moreover, the outlet of the pump is attached with half inch diameter size PVC pipe and how a hypodermic needle may be fitted to this PVC pipe. After verifying different vertical water pumps, available in the market, I concentrated on Tullu vertical water pump as its stem possessed four - five holes to discard excess water; even if we close the PVC pipe, all the water flows out from these holes. The trials revealed that barrel of 5 ml fitted well at distal end of PVC pipe and this led us to publish a research paper using this new technique (60) for perfusing large animals for recovering schistosomes.

I may cite another problem while dealing experiments on schistosomes and this was related to infecting final host with this parasite. It is easy to infect a host with *Fasciola* or amphistome by feeding the infective stage (metacercariae), which is not the case with schistosomes. Here metacercariae are not formed, rather its earlier stage (cercaria) is infective which is capable to penetrate host's skin to develop to sexually mature worms. Though cercariae are able to develop in the host even through oral rout, this is not followed in the experiments due to inconsistent results.

Therefore, different methods were developed to infect laboratory animals with these schistosome cercariae. As mouse is the most common laboratory model, an easy method, in western countries, was developed where mouse is kept in a container and its tail in a test tube filled with cercarial water ; some time is given for penetrating cercariae which is confirmed by presence of only cercarial tails (cercarial body penetrates host's skin) in the water. For infecting larger animals like calf, lamb, a cup like device was developed, again in western countries, which was attached to the skin of the host by creating negative pressure (after adding cercariae) and animal is sedated to hold this device (containing cercarial water) on host's back or flank region.

At IVRI, Izatnagar, India, the laboratory work on schistosomes was started by Dr H D Srivastava and Dr SC Dutt in 1950. For infecting mice by 'tail immersion method' they put whole mouse in a wide mouth glass bottle (500 or 1000 ml size) which was partly filled with cercarial water so that tail and abdomen remained in touch with water. It was easy to leave the mice in such bottles for any time thereby enabling penetration of schistosome cercariae in host's body.

For infecting large animals like piglets and lambs, these workers used plastic tubs or buckets which were filled with cercarial water and animals were allowed to stand in this tub for varying time so that cercariae may penetrate their skin. Some workers followed oral rout or instilling the cercariae into the nostrils of the host.

As same procedure was followed in the two experimental groups, the significant variation in schistosome recovery between the two groups was ascribed due to host species variation or effect of the treatment. Nevertheless, the method was not without fault as was supported by inconsistent fluke recovery from the same group of animals. But there appeared no other alternative simple method hence these methods continued in India for about forty years for infecting large animals (16,60).

I also faced this problem during my work under national fellowship award in 1995 when I have to infect lambs, piglets or goat kids with schistosomes. What may be the best way of infecting large animals with schistosome cercariae that may be considered accurate but simultaneously be less cumbersome ? For simplifying the method, we discarded the complicated western method of attaching cup on host's skin by creating negative pressure and sedating the animal. But neither oral rout nor bucket procedure was accurate for carrying out research program. In short, our problem was to keep the live cercariae in touch with host's skin for at least half an hour under the conditions where they not only remain alive but also biologically active.

During this time, our market was flooded with polythene bags of different sizes. We considered it worth trying to see biological activity of living cercariae in the polythene bags. To our surprise, the schistosome cercariae not only remained alive for hours together but were seemingly quite active in these polythene bags. The next problem was to attach this bag to the host and check if cercariae penetrate its skin and reach to the maturity. For our preliminary work, we selected guinea pig and tied, the two polythene bags (about 5-10 cm size), filled with cercarial water, on their two hind legs or paws, and held it above to the ground to avoid damage to the bags. Checking cercarial water, after half an hour, and finding only cercarial tails confirmed penetration of schistosome cercariae into the host by this 'Polythene-paw' method. But mere penetration was not enough till we could not recover mature schistosomes from these guinea pigs. The animals were sacrificed, after three months of the infection and their examination resulted in recovery of adult blood flukes along with presence of their eggs, thereby confirming success of this new method for infecting the host with schistosome cercariae (60).

Once it was established that schistosome cercariae remain biologically active for a pretty long time in polythene bags, we tried this method for infecting piglets with schistosome cercariae by tying the polythene bag to its tail and ear-pinna of the goats. Development of schistosome infection in these animals was confirmed by excretion of schistosome eggs in their feces and recovering the flukes following perfusion technique (2).

The above two examples are sufficient to explain how we can change the scenario with little efforts and following principles related to the methods. No doubt, further changes may be made in these newly developed two techniques, if efforts are made in this direction.

However, it must be recognized by our future researchers that they should devote their efforts to solve important problems, only then their work will be appreciated. In above examples, there was a real problem of infecting large animals with schistosome cercariae and the existing methods were inaccurate; likewise recovery of the blood flukes by 'cut and soak' method was not only cumbersome but inaccurate as all the schistosomes could not be recovered by this method. Therefore, the problems were genuine and demanded their solution by developing more accurate and simple techniques which may be followed every where with little facilities and efforts. It is doubtful if any further research may bring drastic changes in these newly developed techniques and without these dramatic changes, the researcher will not get much recognition. This directs us to one important fact that the researcher should try to solve a real problem and do not waste time in solving minor ones which are of little significance.

13

Type of Research

There is only one type of scientific research and that is to solve a real problem or to understand the universe and provide a solution - either with the experiments or by putting a hypothesis. However, we have classified the research in various types as per our comfort. But for the employed researchers, it may be a matter of concern if they are working on time bound research schemes as their services may be co-terminus with the project and they may face the problem of un-employment, after termination of the scheme. According to the scholars' activities and type of research being carried out in the country, we can discuss the subject under following topics.

ACCIDENTAL RESEARCH

It is not necessary that you will succeed always in your research within stipulated period; rather you may remain puzzled for months together to find the answer which remain elusive for a long period. During this distressing period you may find suddenly answer of your problem.

The best example, we may cite is that of Archimedes (287 BC-212BC), the Greek mathematician, physicist, engineer, astronomer who has advanced many hypotheses and solved difficult mathematic problems (61). At one time, Archimedes was asked by the king Hiero II of Syracuse to confirm if the golden crown is made of pure gold or if some silver has been added by the goldsmith. The problem was how to calculate volume of the crown, so irregularly shaped, without melting or damaging its figure. He remained puzzled for days together. One day, when he was taking bath in a tub, he observed rise of water level as he immersed himself in the tub. This event resulted in striking an idea to Archimedes that water displaced is proportional to the volume of the matter that is immersed in water. The density of the crown can be determined by dividing its mass by its volume which was equal to the water displaced. And if density of the crown is lower than pure gold, it will suggest adulteration of gold with silver. The thesis proved correct and Archimedes was able to prove that goldsmith indeed has added some silver also, though he was supposed not to do so. However, the point,here, is to mention that this discovery happened accidently and Archimedes was so much excited by his discovery that he ran away on the streets naked crying 'eureka', 'eureka'.

Another good example is of Sir Isaac Newton who started analyzing the event of falling an apple on the ground. This resulted to the discovery of law of gravity. Persons consider this research, though path breaking, as accidental since all of a sudden an idea hit to Newton to know the reason why an apple or any other matter, comes down and does not go above to the sky.

The best example in biological research, is perhaps, discovery of Penicillin in 1928 - the first antibiotic which changed the history of human medicine. Certainly this was an accidental discovery as Dr Alexander Fleming (6th August, 1881 - 11th March 1955) was having no idea of such phenomenon, neither he was working on any anti-biological agent (62). On the other hand, he was working on a bacteria, namely *staphylococcus* sp. which are responsible for infection of wounds, pneumonia and other ailments. Fleming was culturing staphylococci in the Petri-dishes in his laboratory and studying its different properties. One day, a window of the laboratory remained open overnight resulting in contamination of bacterial culture with mold spores which covered bacterial growth in a Petri-dish. He was just discarding this Petri-dish when suddenly he observed under the microscope that the bacteria, surrounding the blue - green fungus, were either dead or dying due to the mold preventing the bacteria of making new cell-wall and reproduction. This led him to believe that the fungus was excreting some biological material which was against the life of the bacteria hence name of anti-biotics was coined to this biological product. Since the fungus belonged to *Penicillium notatum,* the name Penicillin was given to this antibiotic. Fleming tested this material in animals and humans and confirmed its bacterial killing effect without any harm to the host. Even then, the discovery did not attract attention of the scientific world and it remained in the laboratory as mass production of penicillin was difficult. However, beginning of second world war compelled the governments and scientists to search anti-bacterial agents which may be produced in large quantities and may be used against human infections. In this regards, credit goes to Howard Florey, Norman Heatley and Ernst Chain in developing the methods in 1940s for mass production of Penicillin by employing another variety of the mould i.e. *Penicillium chrysogenum* -this was having potential of producing the antibiotic hundred times more than the original mold. This resulted in mass production of Penicillin and its use revolutionized human treatment making it possible to treat cases of gangrene (a common ailment during the war), syphilis,tuberculosis etc which were hitherto incurable and led to many human miseries. Alexander Fleming, with Howard Florey and Ernst Chain were awarded noble prize in medicine in 1945 for discovery of Penicillin (62). This accidental, basic research became so important that with passage of time a new science of antibiotics evolved with discovery of many new antibiotics which, in fact, changed line of treatment of bacterial, viral and many other infections and created a whole new industry of Pharmaceuticals.

We put these discoveries by accident as scientists got the answer all of a sudden. However, these discoveries became possible not by a layman but by a scientist who possessed deep knowledge of the basic fundamentals of science. In some cases, the problem puzzled them for months together prior getting the answer. This reflects complexities of our brain which, unconsciously, was working on the problem hence realized its solution with an event. In case of discovery of Penicillin, Fleming was well versed with bacterial culture and its growth; for this reason, no growth around the fungi drew his attention leading to the suspicion of excretion of anti-bacterial substance by the fungus. Therefore, it may be concluded safely that though these discoveries were accidental, they became possible only because of deep knowledge of the subject by the concerned scientist.

ACADEMIC RESEARCH

This is the type of scientific research being carried out in our universities and institutes by our university teachers and research scholars. The main character of such research is that it is characterless. It incorporates all type of research -may be basic, applied, general or specific; one more feature of such research is that a particular subject is selected by a devoted scientist who contributes immensely to the subject, raises name and fame of the university. But it is irony that in most cases, the research on that subject terminates with retirement of this eminent scientist or teacher. Mostly research is taken on the topics which are not taken by private organizations as they do not provide immediate financial advantages- even some research is considered purely academic as its importance is recognized after pretty long time. Actually, university is the place where important research having long repercussions has been done and it gets its recognition only on type of research work carried by its faculty. It is important to mention that many noble laureates have worked in universities which got recognition by their outstanding research work. These researches have been trend settlers and have paved new direction of applied research which has transformed human life. Many path breaking discoveries have been made in our well recognized universities.

The academic research may be divided into two categories- basic and applied research. The basic research is identified that deals with the subject irrespective of its applications- generally it formulates a thesis or theory; initially, its importance is not realized but subsequent work opens a new field of applications. On the other hand, applied research deals with the aim of its application either for improving the methodology, a new drug for certain ailment or advancement of our knowledge on a particular topic. Generally, our traditional universities are made for conducting basic research while professional institutes are the place for carrying over applied research.

BASIC RESEARCH

At times, it may be difficult to demarcate basic research from applied research but basic research is the mother of investigations for applied research. For instance Mendel's laws (63) propagated that these are the genes placed on the chromosomes in an organism (till that time DNA structure was unknown) which determine its morphological characters or production traits in animals and crops. Looking to expression of morphological characters, Mendel suggested that some may be dominant genes while others are recessive. And a whole subject by the name of Genetics was developed which was again bifurcated in many branches like Plant breeding, animal breeding, genetic disorders etc.The latest is manipulation of genes or biotechnology to create new varieties of the organisms. All this has been possible only with the basic research of Mendel who is regarded father of genetics.

The proof by Jenner (10) that vaccination can protect persons from small pox opened a new field of immunology which expanded vaccination program against almost all bacterial and viral diseases. Since then, a lot of work is carried out on antigens, antibodies, immunodiagnosis, immuno-pathology, autoimmunity, allergies; this has enabled transplant of kidney or liver or other organs by manipulating immune system of the recipient.

The basic research of evolution of species by the Darwin (12) did not appear of applied significance but it opened new areas of research of searching disease nidus, or zoo-geology and alike fields with better understanding of formation of species, subspecies, strains or hybrids among a species.

The basic research is most tedious to perform where a lot of talent, diligence, perseverance is required. Many times, the scientist fails to find answer of his problem or fails to provide experimental proof of his hypothesis but other generation of scientists continue the left over research work and by this way our science progresses.

No doubt, this is the area whose dominance determines progress of a nation. This is the field where most of the noble prizes are awarded. Therefore, it's not surprising that America, the first in world economy, is leading basic research as is witnessed by highest number of noble prizes it is fetching every year. Prior to that Britain was leading the discoveries hence was able to lead the world.

APPLIED RESEARCH

As stated above, it is difficult to demarcate always between basic and applied research. Where will you put the discovery of malaria transmission by mosquito (19) made by Sir Ronald Ross ? Whether it is a basic research proving involvement of an insect in transmission of a disease? Or applied research to find out solution for mode of transmission of malaria. Or where is the place of Jenner's experiment of vaccination against small pox ? There may be debate on the topic, but subsequent identification of role of mosquito in transmission of filariasis, viral infections are categorized under applied research. Likewise, vaccination against cholera or plague is considered as applied research. Perhaps,a fine demarcation is existing between the two. When a research is providing a new hypothesis or a new theory (like insect transmission of pathogens), it is termed basic research and its follow up,using same principles, in other cases is considered applied research.

Leaving this academic debate, there is no doubt that applied research provides immediate benefit to human beings by its results. Thus our green revolution is the result of developing a dwarf wheat variety (64) which tremendously increased wheat production; a cross breeding cattle program is partly responsible for higher milk production in India; white leg horn chicks are responsible for increased egg production or a broiler variety made it possible to produce 2 kg chicken within a month's time; so is the case with many such developments which have tremendously increased our economy. In such studies, already known scientific technology is applied to get new results. Thus the technology may not be new but the end result or variety is much better than the existing one.

This applied research is not confined to development of strains, hybrids of animals or plants but is extended in almost all biological fields.

This may be diagnosis of a disease, or search of a drug, or understanding mode of transmission of an infection. The stem cell therapy, biotechnology, nano-medicine *etc* are the good examples of applied research for searching solutions of so many problems.

No doubt, continuous flow of scientific knowledge across the countries have made it possible to apply new scientific technologies even in developing countries. This has resulted in great improvement in GDP growth of these countries enabling to provide solutions for hunger and disease of its population. You may calculate the progress made within fifty years world wide and all the credit must go to the applied research carried out by these countries.

GOAL ORIENTED RESEARCH

This is the research carried out generally by a team of scientists to achieve a definite result. This is a time consuming research work and carried out by deeply involved and committed scientists. The research goal may be achieved by them within their life time or final result may remain elusive during their life. It does not mean that scientists have failed to solve any problem as they may have solved it partly yet have not reached their target in stipulated time. For instance, vaccine formation against a viral or bacterial infection is easy and there are many vaccines against these infections. On these lines, scientists have planned to develop vaccine against important parasitic infections but have yet not achieved their goal, in spite of devoting life long period. The best example is development of Malaria vaccine which has yet not been commercialized (65). This is because of complexities of life cycle of malaria parasite as well as complex structure of the protozoa.

This type of research is carried out particularly in central research institutes which are meant for specific purpose and to achieve goals in particular field of research. Thus Central Drug Research Institute, Lucknow (under Council of Scientific and Industrial Research) was founded to discover new drug molecules for important infections e.g. kalazar, filaria,amoebiasis, affecting a large human population of the country. A Regional Medical Research Center by Indian Council of Medical Research has been started at Jabalpur, M P to concentrate on tribal health. There are many other RMRC, working in different parts (Dibrugarh, Andaman, Jodhpur, Ahmedabad, Patna) of the country with specific goals. Likewise, Indian Council of Agriculture Research has established many institutes/centers which are concentrating on specific crop (wheat, rice, maize, pulse, coconut, etc) or animals e.g. camel (Bikaner), equine (Hisar), buffalo (Hisar), goat (Makhdoom Dist Mathura), sheep (Avikanagar, Dist

Jaipur) cattle (Karnal), Yak (Arunachal Pradesh), Mithun (Nagaland) and for birds (Izatnagar). In all, there is a chain of research institutes in India working under the aegis of CSIR, ICMR, ICAR on specific topics. Though the research area is confined to particular area/animal/crop, yet the field is very vast as a spectrum of problems are handled related to the particular animal/crop etc. Therefore, research conducted in these institutes may be termed as goal oriented research and scientists are devoting their life long experiences in finding solutions of ever increasing problems. All the details of the institutes may be checked through internet either by opening the website of CSIR, ICMR, ICAR or particular institute/ research center. It is doubtful, how much these institutes are benefitted by isolation from our universities and closing their doors to science students who would have benefitted by handling sophisticated instruments and interacting with our talented scientists.

RESEARCH SCHEMES

There are certain chronic problems that are affecting our country since long and to tackle them, the Government is operating research schemes where scientists are recruited to pay attention on that particular problem. Mostly these are the chronic diseases which are difficult to treat or eradicate hence a whole paraphernalia is established to handle them. The most important research program is that of eradication or control of malaria. We are also having research schemes on tuberculosis, filariasis, leprosy where scientists are working on these topics since long. There are also chronic diseases which are affecting our animal industry like Rinder pest (the reason of establishing veterinary institutes in the countries), food and mouth disease or Ranikhet disease in the poultry. These are the research schemes running from several plan periods and scientists are engaged to contain the ailment and develop new techniques to tackle the disease. Beside these non-plan schemes, there are many short time research schemes which are sanctioned and operated for a limited period of time. These research schemes engage scientists from different fields like zoology, entomology, helminthology, physicians, surgeons and veterinarians. The employed staff has a one chronic problem that they are not sure when the scheme may be terminated and whether their services will be absorbed in parent institute or not; there are many departments which do not have provision of absorbing such displaced persons. For this reason, no serious work is being carried out in such schemes and scientists remain busy to find a permanent job. In summary, working of these research schemes is affected by bureaucratic behavior of the administrators, coupled with financial problems and there is very little new knowledge that is emerging from these research schemes.

Mostly, these research schemes are applying the proven technologies to contain the problem and in that sense are important for the country.

PSEUDO RESEARCH

As the name suggests, this is not research but only its exhibition pretending that the faculty member is engaged in scientific research. This type of work is mostly carried out in government institutes and universities. In recent years, private colleges and hospitals have also joined this category merely to save taxes or to attract students and patients. The faculty members adapt different methods for including their name in research papers, resulted from works of other scholars. A common practice is putting their name as first author in research papers resulting from thesis work of their students - at times, student's name is placed at third or fourth position, when some other important personality becomes interested for inclusion. Although the student murmurs on the injustice made in the hands of his guide, he cannot speak much for fear of his research career.

Another common practice of these pseudo scientists is to insist to include their name in the research papers where any equipment,under their possession, is used or if biological samples are collected from animals/birds/wards under their control. When they have to carry research work or have to allot a research problem to their student, these persons search the journals to find out latest research work being carried out abroad. They repeat this work merely by changing the experimental animal or a drug but without changing plan of the old work. We have witnessed even submission of research schemes to funding agencies, following this tactics. Obviously, such research is carried out not for finding new solution for an old problem but merely to get higher posts or to compete with the faculty members who are deeply engaged in their research.

These pseudo scientists are more worldly creatures and are fully aware about the importance of coming in light in the eyes of the administrators who may be Dean of a college or Director of an institute or vice chancellor of a university. They keep watching daily routine work of the administrators - his time to remain in office, time of visiting different departments, odd times or holidays when administrators may visit libraries, departments, gardens, canteens. And you may see these persons in the library during these times, searching journals or sitting in the laboratory with different equipments etc. It is not that they are sincere to their work as you will not find them during office hours except being attentive in odd hours. Mostly, their disinterest to research is reflected by poor quality and low number of research publications, for which they have various excuses. Irrespective of research outcome, a constant presence of such persons at these places during odd hours provide great impression before the administrator, bringing them in good books and this pseudo researcher may be successful in getting promotion or visiting a foreign country for a scientific cause.

These pseudo researchers become successful, in short time, in getting extra mural responsibilities like library in-charge, hostel warden, chairman/member of different committees etc. Thus, it is easy for them getting selection on higher posts on account of additional extracurricular activities, snatching the just claim of a talented and serious teacher. This resulted in complaining to UGC and other organizations etc that selection committees are biased in selecting the teachers to higher posts. The UGC/ ICAR insisted to prepare a score card and ascribe marks for teaching, research, extension,awards etc so that genuine persons should not be rejected over pseudo scientists. Even after preparing this score card, genuine good research workers, devoting their whole time in research, were denied the promotion while pseudo researchers were successful in getting selected on higher post. The method was simple by keeping lower marks for academic awards, equal marks for extra curricular works like warden-ship, committee work and providing full marks (suppose 20) if a person is having only 10 research papers. Here no emphasis is given to quality of research work, journal where research is published or sequence of authorship- whether the person is first or last author of the paper. The most important fact in this score card is to limit the number of research papers getting full marks and making excess research paper numbers (which will obviously be of serious research worker) redundant. Finally, the high power selection committee has the power to allot marks for personal interview. Intentionally, higher marks, as high as 30 out of 100, are earmarked for this personal interview. Thus it is easy to compensate the deficiency of the blue eyed boy in personal interview category thereby paving the way of selecting pseudo researcher by supposedly in an unbiased manner. To the extreme, a vice chancellor of an Agriculture University and chairman of the selection committee was heard advocating selection of a pseudo scientist on the post of Professor over an award winning scientist before selection committee since the Vice Chancellor has to carry over several non-academic works which can be performed only with the help of such pseudo scientists. As per the Vice Chancellor "sincere scientists are of no use to us as they remain engaged in their research work". Where chairman of a selection committee is of such opinion, how one can prevent selection of these pseudo scientists on higher posts or how our universities can perform world class research ?

Irony it may be, that these pseudo researchers,after retirement, but with material success have also been seen complaining of not achieving anything in their life.

There is now a consolation for real research workers that UGC/ ICAR/ICMR have formulated CAS or career advancement scheme, where these talented scientists will get their promotion within stipulated time, irrespective of liking or otherwise of the administrators.

If you are planning to make research as your career, be aware of such faculty members and always resist to carry research problem under their guidance. When they themselves are lacking any idea, how they can inculcate research ideas to their research scholars ? As per Ravindra Nath Tagore, only a lighting lamp has the power to ignite other lamp and a teacher must be a lighting lamp.

CORPORATE RESEARCH

I have separated this research from research for money due to the reason that the latter research is done by deeply motivated scientists who wish to prove something new to the world and money is secondary which flows as a result of new innovations. In contrast, corporate research is that type of research which is carried out by the corporate houses.

Indeed, the experience in this regards has not been encouraging in India. There is always a cry that our industry should come forward to invest in research and development but till now the result is discouraging.It is observed that these corporate houses carry production of their products using old technologies and whenever a new technology appears necessary, it is borrowed from developed countries,after paying hefty royalties for these so called new technologies. Presently, environment is not conducive for carrying research in the corporate houses; neither there are good equipments nor other facilities are existing. They are engaging staff on minimum wages with lesser qualifications.

All the research carried out by a private company or non government organization may be grouped in this category. If we restrict the topic to biological sciences, this research is carried out by private hospitals, pharmaceutical firms, seed houses, food industry and some research centers, operated by corporate houses. Now, some non-government organizations have opened universities, medical, veterinary, dental and agriculture colleges.The government of India has provided many incentives to augment research in these private institutes; so much so that 125% income tax rebate is admissible on the expenditure incurred on research. But it is a matter of debate whether all these incentives have encouraged corporate research and if the promoters of the company or colleges have ever given a serious thought for conducting good research. A deep study reveals that till now the research has been a tool to get income tax rebate by the stakeholders with little respect to the fruits of good research.

So many facts support my contention. In most of the cities, including metro cities, almost all the private hospitals have added suffix of research merely for getting income tax rebates. These hospitals do not appoint any scientific staff to conduct scientific research or analyzing disease data. Neither these hospitals are developing any advanced surgical or diagnostic technique; rather these are imported and applied in the hospital for financial benefits; at times, their surgeons and physicians are sent to some more advanced hospital of an advanced country to learn a new technique which is applied by the Indian hospital only to fetch extra money for the technique.

The situation of a Pharmaceutical company is not better. No one is interested to operate a full fledged research center which may engage itself in developing new drug molecules for ailments. The main excuse of a drug company against research is exorbitant cost which will incur to develop an effective molecule hence it is not possible for any Indian company to undertake such research. Even no one has advocated multinational firms to open a joint research center in India which will be many times cost effective in comparison to that prevalent in developed country. Earlier, there were some research centers in India, like one operated by Hindustan Ciba and other by Glaxo but both are now closed. Almost all the growth of Pharmaceutical industry in India is depending on generic formulations and modifying the techniques of production to circumspect patent laws.

In recent years, we have witnessed emerging of some private companies which are engaged in stem cell therapy, gene therapy or producing biological products through biotechnologies. Most of these centers appear commercial avenues with main aim of utilizing the existing techniques for commercial purpose. It is still to be observed, how much emphasis these centers will provide for advancing the existing knowledge.

The situation of a private medical, dental, agriculture, veterinary college is still worse than government colleges. Presently, it is elusive that private colleges excel and provide good opportunities for education and research to our young generation, who gets admission in these colleges,after paying a hefty fee. Though there is a vast difference in fee structure between private and government colleges, yet the private colleges have failed to provide even equivalent facilities to their students and staff. Except a few teachers, the staff is not well paid and not as per UGC, AIETC recommendations. Their laboratories are also devoid of required facilities and equipments.

The most wild allegation is that the private institutes are meant for minting money and neither they are interested in imparting world class education nor in encouraging good research. This is the reason, why our industry claims that almost 95% students are unemployable due to lack of training and knowledge. This is a sorry affair of events since the academicians were hoping improvement in education with inclusion of private houses/trusts to education and research; in present scenario, they have disappointed the academic world.

In recent years, a ray of hope has emerged where some renowned anthropologists have donated their life savings for advancement of education or opening of a world class university in the country. As time is changing, we may aspire that corporate houses will understand importance of R & D and more incentives will be given to the researchers for new innovations. It will be beneficial to undertake research, either independently or in joint venture with other centers, so that India may gain an edge over other countries in applying them as new solutions for old ailments. But India has still to wait for large number of entrepreneurs who are willing to join the scientific world not only for profit but for welfare of the society or students.

RESEARCH FOR MONEY

Many of you who are in their late twenties or so might be dreaming to earn money and only money and become the richest person of your state or the country. At the onset, I may mention if you are dreaming only for money, research should not be your aim and you may repent joining it at late hours of your life. This is because you are not having any new dream or new idea but only an illusion to become a billionaire. Here, I may add that you cannot earn money from anywhere, if you are thinking only for money and nothing else. According to an old adage those who chase 'Lakshmi', she never obliges. This implies that money cannot be earned by imagination alone but one has to work hard to earn it; if you are working in right direction, money or Lakshmi will follow you. Further, human history tells us that nobody has found satisfaction or happiness in his life by chasing money as the only goal of his life.

It is not that research has not earned money but it was not the goal of that scientist, rather he was engrossed in his work and money poured in without much efforts. In fact, some of the richest persons of the world became so because of their innovative ideas. You are well aware about the Nobel prize which was instituted in 1895 by the Swedish Chemist Alfred Nobel (21st Oct,1833- 10th Dec 1896) from his earnings (66) on account of discovery of dynamite -to be used for breaking stones but paved the way of development of guns and artillery which altogether changed human history. So is the case with American engineer Henry Ford (30th July, 1863-7th April, 1947) who founded Ford Motor Company at Michigan, USA on 16/6/1903 and sponsored development of assembly line technique of mass production. Presently, the Ford Foundation is responsible for a number of charitable donations and works. There are many other examples and recent ones are those of Bill Gate (USA) and Narain Murty (67) (India), the founders of Microsoft and Infosys information Ltd, respectively. I may emphasize, here, that these two personalities did not have capital to erect a big manufacturing plant for producing steel or aluminum, but it was their intellectual capital which resulted in innovations fetching money, that could not have been earned from anywhere- not even by any steel or aluminum plant. But these personalities did not chase money, instead they concentrated on their research and money came as a byproduct.

Obviously, big money can be earned by launching an industry related to your research activities. If we study history of industries, the early eighteenth and nineteenth centuries had been the era of heavy industries -textiles, steel, auto mobile, petrochemicals and alike one. These industries required huge capital to establish machines and other infrastructure which was not possible for a middle class person hence only offspring of an industrialist was able to establish a new industry. But the twentieth century has changed the scenario where huge capital is not essential for establishing a company and there was emergence of "Venture capitals" who are interested to fund any new idea provided it is worth investing. This is the reason that this period is called Knowledge era with presence of knowledge industries.We have witnessed emerging of new companies which are grouped under service industry. Here, with the invention of computer and related services even a middle class man with little capital can establish his own industry and can be the richest person of the world. This is what Bill Gates and Narayana Murthy demonstrated by founding their own companies, Microsoft and Infosys, in USA and India, respectively.

Even the persons engaged in biological research have been successful in establishing their own companies and have in fact given a run to multinational companies. We may cite here a few examples of pharmaceutical companies in India. Till early eighties, multinational pharmaceutical companies were dominating Indian markets by supplying imported drugs at expensive rates and were earning good profits. This scenario was changed by the efforts of two entrepreneurs - one was Bhai Mohan Singh (68) who established Ranbaxy Pharmaceutical company in Chandigarh, Punjab and other was Dr K Anji Reddy (69) who erected Dr Reddy Laboratories at Hyderabad, Andhra Pradesh. Both have been conferred many national awards including Padma Shri and Padma Bhusan.

Earlier,Ranbaxy was a drug distributor company which was defaulted and taken up by Bhai Mohan Singh (1917-2006) in 1952 for a sum of Rs 2.5 lakh only. The company launched in 60's its own famous brand "Calmpose" which was a copycat of Roche's Valium as this was not patented in India. However, India adopted a regime of process patents in early 70's where the process of manufacturing particular drug could be patented in the country but not the product or the drug (changed this law in the year 2005 as per World Trade Organization to allow patent of the product/drug as well; the patents are for 20 years of innovation of the molecule/Drug and when they lapse, non-patent holders can start producing and selling the drug). This provided a clue to entrepreneurs like Bhai Mohan Singh that one could make any product or drug through reverse engineering and market the same at cheap rate. Exactly this was done by Ranbaxy and later by many other Indian pharmaceutical companies (e.g. Sun Pharma, Cipla, Lupin, Aurbindopharma) who changed the whole scenario by producing and selling generic drugs at lower cost not only in Indian markets but also exporting to other markets. Ranbaxy has become the biggest Indian Pharmaceutical company within a short period and has pioneered in manufacturing antibiotics. (Recently, it is merged with Sun Pharma).

Dr Reddy (1941-2013), a son of turmeric farmer of Guntur district of Andhra Pradesh, completed his Ph.D in chemical engineering from Pune based National Chemical Laboratory and did a job in Indian Drugs and Pharmaceuticals Limited, Hyderabad prior starting his Dr Reddy Laboratories in 1984 with an initial investment of Rs 25 lakh. This company became a two billion dollar plus business within a short period of thirty years and totally changed drug scenario of India. This company transformed Indian bulk drug industry from import dependent in mid-80's to self reliant in 90's and to export oriented industry that it is presently. Dr Reddy Laboratory became first Indian Pharmaceutical company in 1993 to take up drug discovery research in India and was also a first non-Japanese Asian Pharmaceutical company which was able to register its shares in New York Stock Exchange in 2001.

Biotechnology, a new developing branch of Biology, has also tasted its success in business in India and that too in the hands of a woman entrepreneur -Kiran Mazumdar Shaw (70) (Born-23th March, 1953). After finishing her BSc Biology from Bangalore, she joined Ballarat University, Melbourne, Australia and became a qualified Master brewer. Kiran Mazumdar Shaw founded Biocon India in 1978 in collaboration with Biocon Biochemicals Ltd in a garage with two employees and a capital of Rs 10,000 in 1978, but has now reached to more than 6000 crore market capital. The first business success of the company was by extracting an enzyme from papaya and later the company transformed itself from an industrial enzyme company to an integrated biopharmaceutical company. It may interest to those who wish to earn money through research that from a meager capital of Rs ten thousands, Dr Kiran Mazudar Shaw was declared India's richest woman in 2004 with a worth of more than Rs two thousand crores.

Till recently, Pharmaceutical industry was the main source of inspiration where life science persons were engaged and developed their own companies. With changing times, they have dared to venture in some very challenging fields. For instance, Sea 6 Energy, (71) Bangalore has been started by the IIT-Madras students in collaboration with global industrial enzyme maker Novozymes to convert seaweed into ethanol; they are developing same process to generate natural gas. Stempeutics Research, (72) Bangalore is working in the area of stem cell based drugs; they are planning to create living, functional tissues and organs to repair or replace tissues lost due to any reason. Invictus Oncology (73) is founded by a team of researchers from USA, and developing nano-medicine which shrinks tumors by cutting off blood supply and continue to sit on the tumor till it dies. There are many more examples which clearly reveal that new innovations are improving health and wealth of the nation and anyone can join this team of enthusiastic researchers cum entrepreneurs.

There are other biology fields also which have provided good business potentials. I may mention here aqua culture where prawn, crabs or other fresh water or marine creatures are reared under controlled conditions and exported to other countries. There is great scope in horticulture, seed production where research may lead to development of new variety of plants, fruits or seeds; biological fertilizer is the new field having great potentials for research and to establish it as a business unit. Dr B.V. Rao (74) was able to develop a strain of chick which is able to produce more than 260 eggs and a broiler variety which weigh about 2 kg within 60 days; This resulted in establishment of Venkateshwar Hatchery (now Vinky Ltd) in Pune, Maharashtra which is first name in poultry industry. The new recombined technology has great potentials of changing our drug industry and scientists, concentrating in this field, particularly vaccines and immunodiagnosis, may have all chances of great success. Though, some new breeds of sheep, pig have been developed which are good in wool quality or meat production by the efforts of Government research institutes, no one has come forward to use them for establishing a business entity. I have omitted industries like Dairy or animal feed as these are purely business propositions which are being carried out by many entrepreneurs - there old scientific methods are being followed for production and marketing the product with little addition of new knowledge.

As you will note, above examples are taken from real life and are not mere imagination.They represent the efforts of former scientists in establishing their own empire. Here it is also relevant to mention how much hard work they have to put with initial failures prior getting success in their endeavor. However, new developments in India have made the work of future scientists cum entrepreneur easier- they may not have to struggle hard for initial capital as there are many Venture capitals who are interested to lend any amount of money provided your research idea have business potentials. The new patent laws have come enforce which provide a part of future earnings to the scientists capable of developing some new product or its process.

PART-C

14

Searching Literature

When you are assigned with your thesis problem or some problem is puzzling you, the first step is to search the related literature ; here the term literature is used not as it is used in humanities but this means the literature related to the research work published by the earlier scientists on the topic. The searching of literature is a most important and first step in conducting research on a particular topic. This not only acquaint you with the work of other scientists who tried to solve the problem but also confirms the possibility that previous scientists might have solved that problem and yours will be a repeat attempt ; in that case you may tackle other aspects of the problem, if at all they are existing and important in present circumstances.

How one has to proceed to search the literature and what are our sources of searching this literature? Indeed, some part of this discussion comes under library science where students learn about tools of management of information technology by carrying collection, organization, preservation and dissemination of information sources in a systematic manner. The library collection may include books, journals, periodicals, newspapers, manuscripts, films, maps, prints, documents, microfilms, CDs, cassettes, videotapes, DVDs, e-books, audio books, databases etc. It may be a departmental or college or university or a national library–the difference is in space and volume of books and periodicals and other matters present therein. As a research scientist your aim is not to study library science or to become a librarian but to know how you can retrieve the books, journals, periodicals etc from different libraries so that you can extract pertinent information from these sources. Therefore, it is important to know how we can use libraries to its maximum. As you are not expert in library science, it is always advantageous to take help of library assistants in search information resources, particularly if you are visiting a university or national library.

CATALOGUES

Library catalogue or simply catalogue is the oldest method of keeping register of all bibliographic items found in the library or a group of library. This bibliographic item can be related to any information entity (books, journals, reports, theses etc) that is considered library material and each year new additions are made. This system serves two purpose - the first is to confirm what books, periodicals, etc are in possession (though there are stock registers for this purpose) to in-charge of the library and secondly it helps visitors to find a book of his concern. Thus catalogue defines four user tasks-find, identify, select and obtain. For the same book, three catalogues are prepared so that reader may find the book by searching any category of the catalogue. The catalogues are prepared as per author index; title of book ; and subject index.

Dewey Decimal Classification (DDC)

This is the old system of keeping library material where library assigns a DDC number that unambiguously locates a particular subject book at particular place or shelf. Here the material is divided into ten groups and each is assigned 100 numbers with use of decimals so that all materials are classified as per their category.

The main group indicates broader class while decimals reveal sub subjects

000-090 General works

100-190 Philosophy

500-590 Pure science

590 Zoology

591 Vertebrates and 592 Invertebrates

592.123 may be a book on physiology of protozoa and so on....

Universal Decimal Classification (UDC)

DDC has now been replaced with UDC which includes higher features. It can describe any document or object to any desired level of details with identifying book shelves for the particular item.

Here, again catalogues are categorized in two categories

1. ***Author catalogue*** ; The catalogues are arranged alphabetically but by author's name and is helpful when you are sure about the author of the book but not about the title of the book.

An example taken from IVRI library is as follows :

636.708969 Greene, Graig E

56365:R Infectious diseases of the dog and cat

By Graig E Green

3rd Ed, Missouri; Sunders, Elsevier 2006

Xxix 1397 p include appendix and index

ISBN 07216-0062-X

2. ***Title catalogue*** : The catalogues are arranged alphabetically but as per title of the book. Thus it may not be purely on subject basis such as Parasitology, Anatomy, Biochemistry (though some libraries have separated the materials as per such subjects) but as per alphabets of the title (though books/periodicals of specific subjects are arranged together in one shelf to facilitate its searching by the members). Following is the example taken from IVRI library with the start of a thesis title - as post-graduate theses are very important for students and others, good libraries are also maintaining their records as follows :

636 Pathology of experimental aflatoxicosis & aspergillotic

508962430 Pneumonia in broiler chicken.

Thesis

M 687P Mishra,Ajai Kumar

Pathology of experimental aflatoxicosis & aspergillotic

Pneumonia in broiler chicken.

Deemed University, IVRI, MVSc

1.2.95 Izatnagar, IVRI 1995

121 p ; figs, bibliography

001.42 Research Methodology

S23R Sharma, CK

63660 Research methodology/CK Sharma and MK Jain

Shree Publishers and Distributors, NewDelhi

2004, 224 p

ISBN 8188658553

001.4225 Intelligent Data Analysis
B 411I Michael Berthold, David J,Hane
Editors
Springer- Verlag Berlin, 1999
Page 400

These catalogues not only confirm existence of particular book in the library, but also provide other information which help in searching the book in particular shelf in the library. For example, 001.42 informs the shelf where the particular book is available ; 636 is class number, given subject wise while whole 63660 number is given during purchase of the book and is access number. The first letter indicates first letter of the author and second letter shows first letter of the title of the book. Thus, in first case M indicates Mishra and P is the letter taken from title e.g. Pathology. Likewise, in last case, the first letter B is for the author (Berthold M) and second letter is taken from title of the book e.g. Intelligent ...

This UDC code may be given even to a particular research paper that may be searched through Google by logging only this number without knowing name of the journal or title of research paper. For example

UDC code for the research paper "Yeast systematic from phenotype to genotype" published in Journal Food Technology and Biotechnology is 663.12:57.06. You will reach to the research paper through Google by entering merely UDC code of the paper.

Online Public Access Catalogue (OPAC)

This OPAC system was first developed at Ohio State University in 1975 and at Dallas Public Library in 1978. The big libraries are modifying their functioning and are using computer technology in documentation and records management which facilitates to link it to any other library or group of library. Thus OPAC (75) is searching online database of materials held by a library or a group of library ; this method,uses principally a library catalogue, facilitates sorting the material by putting author or title or keyword; you are provided not only details about your book (contents, summary, year of publication etc) but also related books which may interest you. There are ancillary functions informing availability of the specific book on the shelve or issued to a member (also see online library facilities).

DEPARTMENTAL LIBRARY

The first source of finding relevant literature is the departmental library. There are many veterinary colleges in India which are more than fifty years old and have changed their status from constituent colleges of Agriculture Universities to the Veterinary Universities. All these colleges are pursuing research since last fifty years or more and some eminent scientists had led these departments at one or other time. So is the case with zoology departments of the universities,some of which are even hundred years old and their scientists have made pioneer research work. This may not be true for state medical colleges (as they are still following old teaching pattern) though Medical institutes of central government have given emphasis on research and their libraries are rich in scientific material. In recent years, changes are also occurring in medical education with opening of state medical universities and AIIMS in almost every state with fresh emphasis on research in medical sciences.

In the departmental library,the first step is to search copies of theses of post graduate students,reprints of the research papers collected by different scientists ; there are chances of observing scientific abstracts of relevant topics. There may be a list along with reprints of the papers, published by the faculty members of the department -they are of great value as mostly the departments continue to work on the same topic for years together and you may in all probability be allotted a related research problem. Even there may be handwritten notes for understanding new techniques which are important to deal with the new equipments. Research papers apart, the departments maintain annual reports of the work carried out in the department and also technical reports of the research projects ran by the department. There will also be copies of the seminars delivered by the post graduate students and lectures delivered by the staff at different occasions like summer courses, training courses. Search and reading of these materials will be of great help in carrying over your research problem. It is advisable to assort the relevant papers which are needed for your research work and you may proceed further to update the literature from other sources. It is important to mention that certain research materials like post graduate theses, seminars, technical reports of the subject are available mainly in the departmental library and these are important materials which should be consulted prior going to other tertiary material.

The departments are also given grant for purchase of important research or reference books,research journals of their subject; alternatively, college library is transferring old reference books,advances, research journals of specific subjects to the concerned departments. In such case, you will search the literature first in the departmental library prior going to the college/university library for surfing latest publications.

COLLEGE LIBRARY

In fact, this is an important or only source where you may find scientific journals which are required to search full texts of the research papers. It's not possible to collect all the reprints from out sources and one must visit and access college libraries for the journals available. At times, these old libraries might be possessing some important materials- as some colleges are having volumes of British fauna of India, published in early twentieth century or some old journals having important articles or checklists of the parasites. You may also see some complimentary copies of rare publications which may be of great use in understanding subject or history of the subject. Every library has the list of journals,with years, available there and it is of great value for searching the literature in the library.

All the college or university libraries store the text books,reference books, periodicals in separate shelves. Thus all the libraries have separate apartments where journals or periodicals are placed subject wise and chronologically and you may take help of library assistant to search particular shelf of your choice. There may be a separate shelf depicting latest arrivals including books with latest date of publication. Beside the books, you may also observe a booklet,informing new and forthcoming books by the publishers. You may select the book of your choice and request your head of the department to place order for purchase following institute's rules and regulations, in this regards.

UNIVERSITY LIBRARY

In veterinary universities, only veterinary college libraries are existing and the oldest veterinary college may have largest reference materials. Now, every veterinary university has more than one veterinary college and arrangements have been made to link libraries together to facilitate exchange of reference matter.

Libraries are the main source where you will get Advances, reference books, theses, technical reports. Whereas you can burrow text books from library for reading at home ; generally reference materials are not issued to the members and he is supposed to consult them in library itself. How will you check whether particular library is possessing the required book etc or not ?

ON LINE LIBRARIES

As mentioned above, computer technology has changed face of library and important libraries are going for digitalization ; this system will make it easy to search books and journals from any where provided you have username and password to use that facility. This system allows you to put author name or title of the book for retrieval and in many libraries, the retrieved material provides contents of the book with summary of the book so that you may decide its usefulness; simultaneously, it may provide you list of other related books of that subject. Even if you do not know author's name or title of the book, you may search this or similar books by putting keyword i.e. toxoplasma, human protozoa etc.

NATIONAL LIBRARY

Realizing the fact that it is very difficult to provide sufficient grants to every institute to purchase research journals of every field, the Government of India has launched a massive program to join through network, every institute of India so that anyone can surf the site to find the research paper of their interest.

Earlier there were following two national libraries which were fulfilling the need of research demands in their field :

1. National Library of Veterinary Sciences, located at Indian Veterinary Research Institute, Izatnagar, UP.
2. Indian Council of Medical Research, Ansari Nagar, New Delhi.

Both these libraries are maintaining a large number of research journals, reference books, theses, E-books and students and scholars were visiting these places for consultation. Now, they are launching OPAC (on line public access catalogue) and any library from any where may link to library services to check availability of the particular research material and later he can also access it following the prescribed procedures.

Recently, ICAR has launched a massive program of Consortium for E-resources in Agriculture (www.cera.jccc.in) where all state Agriculture Universities (including Veterinary Universities) can browse research journals (roughly 2,800 in number) by titles, subjects etc with the provision to see full papers and may get reprints of important research material.

On the above pattern, UGC has come up to link traditional universities through :

Inflibnet Centre, situated at Gandhinagar, Gujarat (www.inflib net.ac.in)

This information and library network centre is an autonomous inter-university center involved in creating infrastructure for sharing of library and information resources and services among academic and research institutes; it collaborates with Indian university libraries in evolving information environment. Services include-document delivery, bibliographic union databases, library, soul support, walk in users, print archives.There is also a provision to watch online library catalogues of books, theses and journals available in major university libraries in India.

Inflibnet has initiated interlibrary loans and document delivery services from comprehensive collection of subscribed journals; universities can request for articles from the journal holdings of those libraries.

Looking to the developments going on in internet and linking different sources, it can safely be said that there is no problem,now, in accessing available literature on scientific material ; neither there is need to visit national libraries in person for checking the research periodicals. This is a great difference from old days. Now the emphasis should be to analyze the available work done by previous scientists and design the research project in such a way that some new results should come out.

APPROACH FOR RETRIEVING LITERATURE

This is the first step towards solving any research problem. We have not only to search the pertinent information but will have to note down the same for our future reference and this is done either by using reference cards or our computer system.

REFERENCE CARDS

Consulting research journals or searching scientific literature is not sufficient unless details of related research papers are not noted down somewhere so that they may be studied critically prior framing your research plan and referred frequently during the work and also during writing of thesis or research papers. This is done using reference cards or directly on the computer.

Prior popularization of computers and internets, these were the reference cards of about 4x6 inch size of art paper available in big stationary shops, which were used for recording the references. The cards are inscribed with, Author, year, title of paper, Journal 's name, volume,page number and abstract or summary. During search of literature, when you come across an important research paper, you will write its title with name of author(s), journal, year of publication, volume, page numbers and summary of the paper or what is relevant to your research work- this may be methodology or salient findings of the research work.

These reference cards are arranged section wise (materials and methods, results, geographical area, animal species etc) and then chronologically to facilitate their retrieval. The greatest advantage of writing the reference cards is that a student has to read the research paper which acquaints him familiar with the research work- how it was carried out (materials and methods), what were the findings and how the researcher has interpreted his findings in light of previous work (discussion). If you are not reading the research paper and copying the contents, the topic will remain alien and you will not be able to assimilate many important facts about the problem.

By searching the old *almirah*/literature of the department,you may find a whole number of filled reference cards in the personal files of some senior scientists who have worked on topics relevant to your topic.

This practice of "making reference cards" has now been discarded by the students with the advent of internet. But if you are searching references from the hard copy of the abstracting journals, or old journals, either you have to use 'reference cards' or have to note down the details on your laptop. It may be mentioned that most of the journals have been digitalized from recent years to the year of 1973 and you have to search hard copies for referring old literature. This is necessary while dealing taxonomy or morphology of a parasite (as this work has been done mostly prior that period) and to know the finding of previous workers, whose work is still important.

Instead of following a systematic approach, it is a general tendency with the advent of internet that students directly jump to search research papers or literature through Google or other search engines. And when they fail to retrieve research papers on their specific topic, they inform their guide that no work has been done on that particular topic ;this fact only reflects how badly they have started searching literature. Some students, who are more systematic, may go to Wikipedia to know more basics about the subject. To me, this appears a haphazard way of understanding any problem- and this is not the way to acquire full knowledge on the topic.

There is no harm to log in the search engine but there are all chances that you may not understand the problem correctly without understanding basics of the problem. In fact, consulting research papers through search engines is a tertiary level work and is done after searching primary and secondary sources. Again, it is a mistake to consider that all references are available on the internet and its suffice to learn and accumulate literature thru internet (this is different from memberships of the institutes to certain organizations for retrieving scientific information- see below). Please remember that all these search engines, scientific sites provide only limited information and you cannot get total information by this media. Therefore, a different approach is desired to have full knowledge on the subject.

We should start searching the information in a systemic way e.g. there are primary sources, secondary sources and tertiary sources of information and we should proceed one by one for understanding fully the subject.

PRIMARY SOURCE

It is essential to start with Primary source of information which means reading the relevant topic in a standard text book–it may be a text book on Parasitology, Medical Parasitology, Veterinary Parasitology, Helminthology, Protozoology, Entomology or on Invertebrates. (Here, I am not providing names of the text books as almost every year we are witnessing arrival of new text books; nevertheless it will be prudent to consult latest edition of the text book of a good writer). The text book informs you about the basic concept of the problem as text books refer only the facts or what has been accepted in general by scientific community though some text books may also refer pertinent unsolved problems on the topic. It is always advisable to consult books of different subjects on the particular topic. For instance, if you are searching literature on " Drug resistance" in a nematode or protozoa, you may search the topic not only in Text Book on Medical/Veterinary Parasitology but also text books on Pharmacology and Medicine.

This primary source provides you a broader idea about the subject from where you may imagine unsolved problems. If you read the topic "Schistosomes " in the text book written by EJL Soulsby (year 1982) on Helminths, Protozoa and Arthropods (76), the most popular book,consulted by almost all Veterinary colleges of the country,it contains only a few lines on *Schistosoma nasale, S.incognitum,S.spindale* and in describing pathology of animal schistosomiasis, the book uses description of an African schistosome i.e. *Schistosoma matthei*. By reading this topic, it may safely concluded that till 1982, there was little work carried out on Indian schistosomes that can be included in an international text book on Parasitology (However, with passage of time, good work has been done on the subject resulting in publication of a reference book on schistosomes and schistosomiasis in South Asia). This example also provides us the clue that we should consult different editions of the text books to become aware about the progress of the research on particular topic. Moreover, we will get much more information on Indian parasitic diseases in the books written by renowned Indian writers.

SECONDARY SOURCE

The primary information is followed by studying secondary source of information which are reference books, Advances and scientific review articles appending all the relevant references on the topic published on particular subject. In contrast to the text books, this secondary source of information gives details of the matter as it is written in more details by the experts of that particular topic.

ADVANCES IN PARASITOLOGY/BOOKS ON ADVANCE STUDIES

In Parasitology, Advances in Parasitology is an important secondary source of information which publishes most pertinent research topics and is considered of international standard. The Advances in Parasitology are being handled by University of London, London School of Hygiene and Tropical Medicine, Royal Society of Tropical Medicine and Hygiene, and in recent years it has been joined by Natural History Museum, London and published by Academic Press or Oxford University Press, United Kingdom since 1963 and onwards though there were no issues in the year 1966,1981 and 1984. Each year the advances cover important topics in Parasitology written by imminent international scientist and at times they have published more than one volume in one year. These advances deal in more details a topic after taking into consideration all the works conducted by the scientists globally and discuss the merits and demerits of the findings, citing their own opinions and of others with arguments on each part of the problem. Therefore, it is important to read and assimilate matter put up in these articles. If the topic is important or new, and more information has accumulated within subsequent years, the topic is repeated at later time. For instance *in vitro* cultivation procedures for parasitic helminths by Silverman was published in 1965 (77) and was supplemented in 1971 (78). However, a full review on "The immunology of schistosomiasis" by Smithers SR and Terry RJ was published in 1969 volume 7 page 41-93 and again a full review "The immunology of schistosomiasis by same authors was repeated (with advances) in the year 1976 volume 14 page 399-422. Therefore, it is worth searching if your problem has been covered in any volume of these advances. You may consult Advances of Parasitology of the year 2012 volume 79 which provides contents of previous volumes at one place, thereby knowing if your topic has been covered till that time- of course, you should search new advances also for confirming presence of the topic in any volume of the advances.

It may happen, that your topic may also evince interest in related subjects. For instance, "Recent advances in the anthelminthic treatment of domestic animals" has been covered by Advances in Parasitology (year 1964) (79) while Advances in Chemotherapy Year 1968 has dealt chemotherapy of Trichomoniasis, chemotherapy of cestodes infections and chemotherapy of trematode infections, separately (80). Or a topic on Theileriosis has been covered both by Advances in Parasitology and Advances in Veterinary Medicine. Obviously, it is always advantageous to consult advances of other subjects (e.g. advances in drug research, advances in immunology, advances in molecular and cell biology), as well since these Advances have different priorities and may incorporate different references on same topic, highlighting different aspects.

REFERENCE BOOKS

The other Secondary source of reference is our reference books which deal the particular topic in greater details. These reference books may cover the whole subject of Parasitology or Helminthology etc with chapters on different topics,written by the experts, providing latest information with references. Obviously most of the reference books are written by multiple authors and edited by either one or more scientists.

BOOKS WITH INDIAN CONTEXT

Some good reference books with Indian perspectives have been published in pre-independent India. Patton WS and Cragg FW (1913) published " Text book of Medical Entomology". A voluminous supplement (370 pages, Vol 10) of Indian Journal of Medical Research was published by Sewell RBS in 1922 on "Cercariae indicae" providing all the information on existing cercariae of the country. The series of publication of " Fauna of British India including Ceylon and Burma" are still considered classical publication. Though whole series is important for persons dealing different aspects of biology, following volumes have covered organisms related to parasites :

Southwell, T 1930 The Fauna of British India including Ceylon and Burma. Cestode Vol I and II Taylor and Francis, London.

Baylis, H A 1936 & 1939 The Fauna of British India including Ceylon and Burma. Nematoda. Vol I and II Taylor and Francis, London.

Bhatia B L 1936 The Fauna of British India including Ceylon and Burma. Protozoa. Taylor and Francis, London.

Seniorwhite R, Aupertin D and Smart J 1940. The Fauna of British India including Ceylon and Burma. Diptera - Calliphoridae Taylor and Francis, London.

I will like to mention here the book of G D Bhalerao (1935) "Helminth Parasites of domesticated animals in India "(ICAR publication) which provided important information on the helminth parasites of livestock existing in India along with their intermediate hosts ; however, many of prescribed observations are not tenable in light of subsequent research work, and many times it may be confusing to the young Parasitologists. Therefore, whenever such old books are consulted,a great care is needed to interpret the subject as ensuing researches have disputed many of these old findings.

In Post-independent India, there are not many reference books covering Indian Parasitology but they are worth studying as they are written by the experts of their subject. For example, the book of Sen SK and Fletcher TB (1962) " Veterinary Entomology and Acarology for India " (ICAR publication) is important for knowing existence and life cycle of insects,ticks and mites of Veterinary importance in India. The book "Review of Parasitic zoonoses" (ISBN 81-85386-21-8) edited by SC Pariza,Head of Microbiology, JIPMER, Pondicherry, published (AITBS Publishers and Distributors, Delhi) in 1990 deals all parasitic zoonotic diseases taking into consideration zoonotic aspects and Indian work. Likewise, " Text book of Helminthology" (ISBN 9788171640058) authored by Ku Suresh Singh, former Head, Division of Parasitology, IVRI, published by ICAR in 2003 (though contain references only up-to 1975) is important for knowing helminthes existing in India. Some other important reference books in Indian contest are as follows:

1. Parasitic diseases of buffalo in India. Bhatia BB and Chauhan PPS 1984, Cosmo Publication, Delhi. Page 118 (there are considerable printing errors in the book).
2. Hand book : Fresh Water Molluscs of India. NV SubbaRao, Zoological Survey of India.Calcutta. 1989. Page 289. This is a very useful guide with original photographs or line sketch for knowing fresh water molluscs of India.
3. Helminthology : Editor N Chowdhury and Tadda I, Narosa Publishing House, New Delhi 1994. Page 373. There is a chapter by N Chowdhury on existing helminthes in Indian subcontinent beside basic topics on helminthes.

4. Helminthology in India : Editor ML, Sood. International Book Distributors, New Delhi 2003. Page 617 (ISBN 9788170892939). The book contains important invited topics from Indian authors related to Helminthology in India.
5. Infectious diseases of domestic animals and zoonoses in India, Editor Veena Tandon and BN Dhawan. The National Academy of Sciences India, Allahabad 2005, Page 296. This special issue of Proceedings of National Academy of Sciences, India has discussed important parasitic, bacterial and viral infections.
6. Schistosomes and schistosomiasis in South Asia,MC Agrawal, Springer Pvt Ltd New Delhi 2012 Page 351 (ISBN 9788132205388). This reference book deals all the details of schistosome species, existing in Indian continent and dealing the subject of South Asia which is known for having different schistosome species.

ICAR : Some reference books are in the form of a monograph dealing specific subject; mostly these monographs are written by the scientist who has devoted his life on that topic hence is expert on the particular topic. It is pertinent to mention that ICAR, New Delhi has published some important and beautiful monographs related to Parasitology. These are :

- Srivastava HD and Dutt, SC 1962 Studies on *Schistosoma indicum* Research Series No. 34. Page 91.
- Deo P G 1964. Round Worms of Poultry.
- Gill BS 1977 Trypanosomes and Trypanosomiasis in Indian Livestock Page 137 (again revised by Gill in 1991).
- Geeverghese G and Dhanda V 1987. The Indian *Hyalomma* Ticks (Ixodoidea; Ixodidae) Page 119.

ICMR : The Indian Council of Medical Research, New Delhi have also published some important books which may prove as a guide for carrying out biological research in India :

- Ganguly NK, Jotwani G, Mathur R and Valiathan MS, 2008. Ethical Guidelines for Biomedical Research on Human Participants. ICMR, New Delhi.
- Satyavati GV, Gupta AK, Tandon N and Seth SD 1987. Medicinal Plants of India. Vol 2. ICMR.
- Reviews on Indian Medicinal Plants 2004 to 2013. Vol 1 to 12 ICMR.

- Indian Medicinal Plants in the Management of Lymphatic Filariasis. 2012. ICMR.
- Emerging and Reemerging infections. 1996 Special issue of Indian Journal of Medical Research.

OTHER PUBLISHERS

There are monographs published by other publishers but dealing Indian parasites.

Dutt SC 1980. Paramphistomes and Paramphistomiasis of domestic ruminants in India. Punjab Agriculture University, Ludhiana. Page 189.

Mandal AK 1980. Coccidia and coccidiosis of poultry and farm animals of India. Zoological Survey of India. Tech Mono no. 5. Page 87.

Mandal AK 1987. Fauna of India: Protozoa -sporozoa : Eucoccidiida: Eimeriidae. Zoological Survey of India. Page 460.

Gupta NK, 1993. Amphistomes: Systamatics and Biology. Oxford and IBH Pub Co Pvt Ltd, New Delhi. Page 696.

Singh G and Prabhakar S (editors) 2002. *Taenia solium* cysticercosis: From basic to clinical science. Page 480 (ISBN 9780851996288). Oxford University Press, New Delhi.

Geeverghese G and Mishra AC 2011 *Haemaphysalis* Ticks of India. Academic Press (ISBN 9780123878113).

Above monographs must be consulted for getting overall knowledge on the subject in Indian perspectives.

INTERNATIONAL PUBLISHERS

There are chances of absence of any reference book/monograph written by Indian scientist on the topic where you are planning to work. In such cases, it must be understood that though scientists are dealing the subject with relation to their country yet it cannot be restricted to geographical boundaries therefore it is always advantageous to consult international reference books, whenever available on the desired topics.

Some important reference books/monographs published by international publishers are given below ; these are important as there appears no such work from Indian scientists.

Important helminth infections of South East Asia (Diversity and potential for control and elimination) Part A.2010 (Editors Zhou, Bergquist, Olveda and Utzinger) Advances in Parasitology Volume 72 Page 480. ISBN 9780123815132.

Important helminth infections of South East Asia (Diversity and potential for control and elimination) Part B.2010 (Editors Zhou, Bergquist, Olveda and Utzinger) Advances in Parasitology Volume 73.

Conor R Caffrey (editor), 2012 Parasitic Helminths. Paul M Selzer (series editor). Wiley Blackwell Page 540 (ISBN 9783527330591) This is 3rd volume of the series and dealing in drug and vaccine development against helminths.

Sibat HF (editor) 2013 Novel aspects on cysticercosis and neurocysticercosis. Intech ISBN 978-953-51-0956-3 Page 364 (This is an open access book (www.intechopen.com/subjects/medicine) with DOI:10.5772/33".

Littlewood DTJ, Baker JR Muller R and Rollinson D (editors) 2003 The Evolution of parasitism : The evolution of parasitology vol 54 Phylogenetic perspective In Advances in Parasitology. Academic Press Inc London, UK page 416 ISBN 0120317540.

Baker JR, Muller R, Rollinson D, Hay and Randolph Rogers (editors) 2000. Remote sensing and geographical information systems in epidemiology. Vol 47. Advances in Parasitology. Academic Press Inc London UK page 357.

J.Webster (editor) 2009. Natural History of Host Parasite Interactions. Vol 68 Advances in Parasitology.

Bechage NF and Drezen JM 2011. Parasitoid Viruses, Symbionts and Pathogens. Academic Press USA Page 312 ISBN 9780123848581.

Ronald Fayer and Lihua Xiao (Editors) 2007. *Cryptosporidium* and Cryptosporidiosis. Second Edition. CRC Press, Atlanta USA Page 576 ISBN 1420052268.

David S Lindsay and Louis M Weiss (editors)2004. Opportunistic infections. *Toxoplasma, Sarcocystis* and *Microsporidia*. Springer USA Page 245. ISBN 978-1-4020-7846-0.

Peter J Myler and Fasel Nicolas (editors) 2008. *Leishmania;* After the Genome. Academic Press UK Page 306 ISBN9781904455288.

CABI, UK have published important monographs on parasitic diseases of international importance and should be consulted for getting an holistic view on the topic.

Springer Link has published book series under WORLD CLASS PARASITES which cover many important parasites and one may look the contents by logging link.springer.com/search?acet-series (there are books on trypanosomes, theileria, leishmania, filarial, geohelminthes etc published between 2002-2007).

REVIEW PAPERS

Many research journals, including abstracting journals, invite imminent scientists to write a review article on the research work where he has devoted sufficient time and which need attention of larger scientific community to advance future work. In recent past, some Indian scientists have published important review papers in international and national journals discussing the subject in Indian perspective. These review articles are very important for the research workers as they provide discussion of a related topic by an imminent scientist. These review articles should be studied grasping the arguments and areas where you are differing and like to pursue the work to prove otherwise. Moreover, this is the source where you will get many cross references which may interest you.

However, a disadvantage, always associated with these reference books, monographs, review papers, is that the concepts, promoted there of, may have become outdated or would have been replaced by new hypotheses. Even the prevalence rate of a parasitic disease alters by changes in ecological or sociological characters. Therefore, please do not forget to know the latest information by consulting pertinent research papers, published in ensuing time.

TERTIARY SOURCE

In the primary and secondary source of information, in the form of text books or reference books, the author deals the subject in a more holistic way, considering the opinions of other scientists and putting his opinions as well. In contrast, tertiary source of information is the research papers (full length or short notes) published by one or many scientists together, after conducting research on that particular topic. The tertiary source of information is related to the scientific journals which publish research articles on the particular topic. These scientific journals are categorized subject wise (Parasitology will have many scientific journals depending on sub-subjects), contribution wise being international, national and state level journals- now they are also categorized on the basis of impact points, how much the research journal is referred by scientific community internationally. Though research papers are main source of tertiary information, theses, project reports, status reports etc are other useful sources for searching pertinent information.

These research papers are important as here the scientist is dealing a particular problem and inform scientific community his methodology or materials and methods of his research work and his results in greater details. The scientist has to discuss how and why his results are important and why he observed differences from others' results. Thus reading a research paper tells many things to another researcher. One has to critically observe what materials have been used and what were the techniques employed in such research work (it may happen that your laboratory may not have such facilities or vice versa). It also provides a window to analyze what results the scientist has recorded and why; this gives you a chance to find shortcomings in materials and methods of the scientist or irrationality in reaching conclusions. It is also important to note that the researchers are permitted to restrict their comments in discussion only to the work carried out by them. Thus, studying a research paper provides basic back ground to understand how another scientist has dealt the same subject and why and how he received these results ; it also provides you idea to think how you may get different results by changing your methodology.

However, there is a systematic way of searching research papers for excluding the possibility of not consulting important journals and research papers on pertinent subject.Nevertheless, there are hundreds of research journals,published internationally, and it is not possible by any library to subscribe all these journals. Moreover, consulting each and every research journal will be a herculean task for an individual. To facilitate the research workers, therefore, there are abstracting journals which edit a large number of research journals of particular discipline (e.g. Helminthological Abstracts) and provide abstracts and other details of all the research papers (including short notes, technical reports, annual reports), appeared in these journals.

YOUR APPROACH FOR TERTIARY INFORMATION

An important question is what should be your approach in accessing tertiary information i.e. first searching abstract journals and later access research journals or directly reach to research journals.

In my opinion, it is advantageous to search literature first in abstracting journals by logging key words of your choice, you will be able to see all the research papers related to your subject and published in any journal (that is covered by Abstracting journal but may not be available in your library). Thus there are little chances of missing any research paper published in any of the journals including international ones. After checking abstracting journals, you may search the research journals where these papers have appeared. If you vet directly research journals, it will be more tedious to check such a large number of research journals; additionally, you will not be able to check some important research journals due to their not being subscribed by your library hence you will remain unaware about such important research papers that may be seen in abstracting journals.

Therefore, in my opinion, a researcher should start by consulting abstracting journals and note down important research papers which may be consulted as full paper by reaching to that specific research journal. However, there are many abstracting journals which are catering to Parasitology- some are common to the Parasitologists while others are more specific to Veterinary/Medical and Basic Parasitologists. Some important abstracting journals are given below :

ABSTRACTING JOURNALS

The abstracting journals are published monthly or quarterly or biannually, which scan almost all the journals of international and national importance on the subject and publish abstracts of the research papers with additional details. The details include authors of the research paper, their official and e-mail address (you may make correspondence on that address), journal's name, year, volume, issue number with page number, publisher of the journal, language of the paper, and also reference numbers used in the research paper with summary or abstract (though the original paper may be in some foreign language, the abstract is provided in English language). These abstracting journals, generally once in a year, also provide list of the journals which are used for abstracting the research articles hence one may confirm which journals have been excluded from abstracting. Presently, almost all the publishers have shifted to publishing the abstracting journals in soft copy format as they are easy to use but mostly digitalization of the abstracting is restricted to 1973 hence one has to depend on hard copy for searching past research work.

Helminthological Abstracts (Series A)

This is the oldest monthly abstracting journal,related to Parasitology, started in 1933 by CABI, UK (www.cabi.org) and began online publication too since 1973. The series A deals with the helminthes of animals while Series B deals with the plant helminthes.

The abstracts are grouped in different categories - Trematodes, Nematodes, cestodes, acanthocephalan (parasitic in man, domestic and wild animals), taxonomy, pathology, immunology, diagnosis, epidemiology, treatment, control, Techniques etc Additionally these journals also publish abstracts of departmental reports, government reports and review articles on the relevant topics. At the last, there is subject and author index which facilitate to locate all the relevant abstracts. These journals are providing annual index also either in December or January issue and it is advantageous to search the abstracts from annual index rather than monthly index. The abstracting journal is adding more than 5,000 new records, each year.

Protozoological Abstracts

As there was no abstracting journal on Parasitic protozoa, CABI, UK launched this abstracting journal since 1956 first as biannual and later to bimonthly publication. It provides information on all aspects of parasitic protozoa including molecular genetics, biochemistry, physiology, morphology, taxonomy, life cycle, immunology, pathology, transmission, epidemiology, diagnosis, treatment and control of protista of man, domestic and wild animals. There is a back file of about 2,00,000 data since 1973 with addition of about 8000 new records, each year.

Review of Medical and Veterinary Entomology

This is again an abstracting monthly journal of CABI, UK which incorporates abstracts covering insects and other arthropods that are important to medical and veterinary sciences. The back file since 1973 possesses 2,50,000 data with addition of 12,000 new records each year.

The above three abstracts are specific to Parasitology and many colleges might not be subscribing them due to financial constrains. Moreover, almost all subjects are having their own abstracting journals (nutrition, animal breeding,dairy science etc) and its very difficult by any institute to subscribe abstracting journals of every subject. Therefore, most of the colleges are subscribing the abstracting journal which contains abstracts of most of the subjects. Some multi-subject abstracting journals are as follows :

Veterinary bulletin

It shows abstracts related to veterinary science. As the bulletin covers a large number of veterinary subjects, not all abstracts of individual subject are covered but it contains only limited abstracts particularly those which are important in veterinary medicine or are of applied nature. Therefore, you will not notice abstracts of the papers related to morphology,life cycle or alike topics which have little relevance to applied veterinary medicine.

Vet CD (CD-ROM Data bases)

This is a type of CD-ROM data base and supplied in the form of a compact disc which is operated with the use of operating system CD. Unlike to Veterinary bulletin, it is abstracting a large number of journals and we observed its just equal to that of specific abstracting journals. This is published by CAB, UK and a subscriber is presented all the copies of VET CD starting from the year 1973 till the date and continue to supplement the information till subscription is made or renewed.

How will you use this CD for searching literature for your research work ? One important fact is that these CD are available for different periods - research papers published between 1973-1988, 1988-1998, 1999-2009, 2009-2012. As These Publishers up-date their contents generally quarterly, you will find new CD containing abstracts on quarter basis. As the contents of the CD suggest, you may use CD of the periods for which you are desirous of getting the abstracts. However, it should be remembered that these CD cannot be operated on any computer unless you have basic soft ware CD which enables its operation on that computer.

How you should search your references from the CD ? After log in the relevant CD, you may type the subject in search column but without putting any adjectives. Remember that some subjects may inform you about thousands of references on that topic hence you should start narrowing down your search. For example, if we search schistosomes, there will be thousands of the references on the topic - so will be the case with search of Malaria or *Plasmodium*. Therefore, we should narrow down the subject such as *Schistosoma indicum*, *Schistosoma* immunity or Schistosomes as per geography - India, Pakistan, Bhutan. There is no harm to search your subject with as many key words as possible, as this gives you chance to locate references but in this process you will get some repeat references at every search. You select desired references of your choice and mark them so that they may be down loaded either in a pen derive or may be saved in words document or may be e-mailed to desired address.

When you are using computer for reference work, it's easy to make different files with different sub-topics ; it may be morphology, life cycle, prevalence in intermediate host / final host, diagnosis treatment etc You may arrange the references chronologically as well as alphabetically as these are the procedures followed in writing any research paper/thesis.

While using computers for your research work, a word of caution is not to depend totally on your computer or CD as at times virus may destroy these files making you helpless at last moment. Therefore, it is advisable to have a hard copy of all the relevant material or e-mail the material to your e-mail address where you may store safely the material under a folder of your e-mail.

BASIC PARASITOLOGY

All the above three Parasitological abstracts are equally important to those who are working on basic parasitology. Additionally, following abstract journals may be consulted by them:

Biological abstracts

The EBSCO information services, Ipswich, Massasachute, USA is publishing this journal covering life science and biomedical research literature published from more than 4,300 journals internationally. There are 13.2 million records since 1969 with addition of more than 3,70,000 citations,each year.

Entomology Abstracts

This is published monthly by Cambridge Scientific Abstracts, UK (www.csa.com/factssheets/entomology) since 1982 and incorporates taxonomy, phylogeny, morphology, physiology, anatomy,biochemistry, genetics, reproduction,biology etc of insects, arachnids, myriapods, onychophorans, and terrestrial isopods. This is more useful to basic Parasitologists.

TAXONOMIC WORK

An important work of Basic Parasitologists is to identify correctly the parasite species which is encountered. This species may be a new record for the geographical area or host species or organ affected or may be an altogether new species. This is the taxonomic work and requires specialized services from a taxonomist. However, a Veterinary or Medical Parasitologist may also encountere parasite species whose identification is must prior proceeding for any other work. The first step should be to know what parasite species of that group are existing in the country ; there are check lists of the helminths published by the scientists i.e VS Alwar (81), HD Srivastava (82) and Thappar (83) that informs about the helminths (as per my knowledge, there is no checklist of protozoan parasites of India) recovered from India and from a particular host species; its always advantageous to have a copy of the same for verification or otherwise (I am not mentioning about "Fauna of British India" as there are many disputed reports hence requires a thorough analysis).

When ever a taxonomic work is undertaken on the parasites, it is essential to consult a standard reference book that discuss zoological keys and various photographs of the parasites in different conditions. The classical books have been several volumes compiled by Yamaguti (84). As it becomes an old publication, CABI, UK has come forward with several volumes which are helpful in taxonomic work and I am giving them for further consultation in respective fields :

Gibson DI, Jones A and Bray RA (editors) 2002. Key to the Trematodes. Vol 1. CABI, UK Page 544. ISBN 9780851995472

Jones A,Bray RA, Gibson DI (editors) 2005. Key to the Trematodes. Vol 2. CABI, UK Page 768. ISBN 9780851995878

Bray RA, Gibson DI and Jones A (editors) 2008. Key to the Trematodes. Vol 3. CABI,UK Page 848. ISBN 9780851995885

Khalil LF, Jones A and Bray RA (editors) 1994. Keys to the Cestode Parasites of Vertebrates CABI UK Page 768. ISBN 9780851988795

Anderson RC, Chabaud AG and Willmot S 2009. Keys to the Nematode Parasites of Vertebrates. CABI UK page 480 ISBN 9781845935726 (this is concise volume of the 10 part CIH keys to the nematode parasites of vertebrates which was published between 1974 to 1983- the new volume is refreshed and reformatted)

Gibbons LM 2009. Keys to the nematode parasites of vertebrates, supplementary volume. CABI, UK Page 424. ISBN 9781845935719

However, it must be mentioned that many traditional universities do not work on Parasitology,like Rani Durgavati Vishwa Vidyalaya, Jabalpur (Ph.D. theses on parasitological problems are of those scholars who have worked mainly at Jabalpur Veterinary College) ; their library is also poor on parasitological literature, neither there is any faculty working on parasitological topic. Therefore, it is not advisable for research scholars to join such institutes to work on parasitological problems. In India, there are only few traditional universities, like Aligarh University, Lucknow university, Allahabad university, Andhra University, Vishakhapatnam where faculty has worked on parasitological problems and it will be prudent to inquire if still some faculty member is working on such problems. Indeed, pure basic work on parasitology are extensively dealt by European and American institutes and it will be better to join them for pursuing research in basic parasitology.

MEDICAL PARASITOLOGY

Medical Parasitologists work on different aspects of the subject hence they should consult above abstracting journals, depending on their research project. However, these journals may not include pure medical cases hence they should supplement their search with an informative system which records details of human cases and in that sense Pubmed is the best source for retrieval of such literature.

PUBMED

This is an important publication of US National Library of Medicine, Bethesda, MD, USA and comprises more than 23 million citations for biomedical literature from MEDLINE, life science journals and online books. However, it does not abstract exclusively on parasitic diseases but from the field of biomedicine, health, behavioral sciences, chemical sciences and bioengineering. Unlike to many other publishers, the access of Pubmed is free to any one interested to look into the abstracts, present on their website. You can log in www.ncbi.nlm.nih.gov for getting all the details and retrieving references of your interest.

Tropical diseases Bulletin

As the name suggests this is the abstracting journal of CABI,UK which provides abstracts related to tropical diseases of human beings. Thus it does not abstract only papers related to parasitic diseases but also include tropical viral, bacterial and fungal diseases.

Indian Science Abstract

As the name suggests it includes abstracts only of Indian publications but of both category- physical and biological sciences. This is a monthly abstracting journal from INSDAC.

The Indian Animal Sciences Abstracts

In recent years, Directorate of information and publications of Agriculture, ICAR New Delhi started publishing half yearly abstracting journal (ISSN 0973-0222) for the help of Indian researchers and contain abstracts only of Indian journals. The headings of abstracting papers are not formal hence you will not see the term "Parasitology" or "Pathology"; instead they are referred in "Pests of animals", "Animal diseases" etc. At last of each number, there is Author index and a Keyword index (not subject index) which facilitates to search the abstracts from key words. Additionally, the abstracting journal facilitates to lodge a request to "The information systems officer, Agricultural Research information centre, DIPA, Krishi Anusandhan Bhavan, Pusa, New Delhi 110012 "(E-mail <hansraj@icar. org.in) for procuring full text of the document,free of cost.

Thesis Abstracts

This is a quarterly publication from Haryana Agriculture University, Hisar and provides MVSc and Ph.D theses abstracts of those which are submitted to Agricultural Universities of India. Beside summary of the work, it also provides information about the student, guide, year of submission etc.

INDEX JOURNALS

These journals were popular prior invention of internet system. The journal provides information about authors, address for correspondence, title of article, year of publication, name of journal, volume, issue number, page numbers and covers a large number of national and international journals. Importantly non of these Index journals provide abstract of the article but made you aware about latest publication in short time from all over world which was not possible by searching the journals in the library. Another important use of the journal was for sending request to the authors for dispatching reprints of the research paper as those time the journals were issuing some free reprints to each author (extra may be purchased) and these were distributed free of cost to the fellow scientists. Due to limited budget, the libraries were subscribing one or two index journals being affordable rather then subscribing some international journals. Some important index journals were :

a) Current contents (life sciences series). This is most popular weekly publication.

b) Index Medicus : This is a quarterly publication mainly covering medical literature.

c) Index Veterinarius : This is a quarterly publication from CABI and covers veterinary subjects.

RESEARCH JOURNALS

This is the basic source where one finds research papers of the subject. If you wish to make your career in Parasitology in India, it is must that you should develop an idea what is going on in research in Parasitology in India and there is no other better way than reading the scientific papers published in parasitological journals originated from India. Equally, it is important to read international journals of Parasitology to acquaint yourself on what is going on internationally. Then only you will be equipped to know what more is to be done so that your work may be appreciated by fellow Parasitologists.

It is important for the students to read full papers of the relevant topics as many as possible and these papers are found only in the research journals. Without reading full papers and depending totally on the abstracts of the papers will not give a holistic view of the subject to a student.

INDIAN PARASITOLOGY JOURNALS

In India there are presently three Parasitology associations which are publishing three important research journals on Parasitology-all on half year basis. The oldest association is Indian Society of Parasitology (established in 1973) which is publishing Indian journal of Parasitology since 1977 though in 1994 the journal is re-christened as Journal of Parasitic Diseases or JPD (ISSN 0253-7168) and being published by Springer India Pvt Ltd, New Delhi where Dr Veena Tandon is acting its Chief Editor (this journal has become online hence may be viewed on the internet). You will find research papers, related to Parasitology from all three disciplines of Parasitology-Zoology,Veterinary and Medicine.

The other is Journal of Veterinary Parasitology - being published from 1987 by Indian Association for Advancement of Veterinary Parasitology (IAAVP). As the name suggests it deals mainly veterinary aspect of parasitology and you will find research papers pertaining to domestic and wild animals including birds. Presently, the journal is being published from Division of Parasitology, Indian Veterinary Research Institute, Izatnagar under Dr P S Banerjee as Editor in Chief and shortly going to be online also.

Since the year 2005, Indian Academy of Tropical Parasitology has been started under chairmanship of DR SC Parija, Head of Microbiology, JIPMER, Pondecherry. The members of the academy are mainly from Medical field those who are working on parasitic diseases. The Academy is publishing on line Journal of Tropical Parasitology and contains mainly articles from medical fraternity.

The National Institute of Malaria Research, New Delhi was publishing Indian Journal of Malariology which was suspended between 1964 to 1980. Since 2002 scope of the journal has been widened to include other vector born diseases hence it has now been named as Journal of Vector borne Diseases (ISSN 0972-9062).

OTHER INDIAN JOURNALS

There are other research journals also in India which publish research work in Parasitology. Some important ones are Current Science, Bangalore, Indian Veterinary Journal since 1921 from Chennai, Indian Journal of Animal Sciences (earlier Indian Journal of Veterinary, science and Animal Husbandry) (a ICAR publication), Indian Journal of Experimental Biology (a CSIR publication), Indian Journal of Medical Research (a ICMR publication) Journal of Indian Association for communicable diseases (ISSN 0254-3575), Journal of Zoological Society of India. There are other journals which are irregular in publication or has stopped its publication like Indian Journal of Helminthology or Indian Journal of Applied Biology and Parasitology. Beside these national level journals, almost every Agriculture University is publishing its own research journal and there are state veterinary/medical associations which are publishing their own journal-generally for the benefit of health practitioners. As now every journal is being accessed by IMPACT INDEX or impact points, generally reputed scientists are publishing their work only in journals which are having good impact index ;obviously, these are national, peer reviewed journals which occupy these positions.

The above information led to an important fact that all the parasitological work, carried out in India prior 1977 have to be searched out in other journals than JPD or JVP. Likewise, the papers which deal more specific parasitological work like taxonomy, are now a days not getting space in IVJ or Indian Journal of Animal Sciences or Indian Journal of Medical Research. It is more time saving to search annual subject index rather than monthly index.

INTERNATIONAL JOURNALS ON PARASITOLOGY

With out accessing international journals of your subject, you can not know what is going on internationally in scientific world. Therefore, it is necessary to search important international journals on parasitology or at least the journals which cover your field of work. Mostly, these are the Academies or Societies which are publishing reputed journals on parasitology and some of them are given below :

- Journal of Parasitology
- Journal of Experimental Parasitology

- Journal of Helminthology
- Journal of Protozoology
- Parasitology Today
- Parasitology
- International Journal for Parasitology

Beside these journals which inform almost all aspects of parasitology, more specific journals have emerged which deal particular fields in more details. A few of them are as follows (the year and volume in that year reveals how old is the publication) :

- Systematic Parasitology (Vol 1 in 1980)
- Journal of Molluscan Studies (60 vol in 1994)
- Molecular and Biochemical Parasitology (85 vol in 1997)
- Parasite Immunology (17 vol in 1995)
- Journal of Invertebrate Pathology (56 vol in 1990)

If you access these journals, one fact emerges out significantly that these journals have reported many events, facts first time and that have been followed by other workers,later on. Another fact is that location of these journals is either Europe or America and both these geographies are devoid of tropical parasitology ;therefore, the main emphasis of research in these laboratories, and journals is related to basic research on parasitology - this includes biochemical studies, immunodiagnosis, vaccinology, phylogenetics etc You will rarely find epidemiological research papers and those which appear are related to joint collaborations either from African or Asian scientists.

All the above journals have their own website and you may directly log in their website where you can check contents of each issue of the journal. If you desire to procure a reprint of paper, published in any of these journals, it will cost you dearly hence it's better to log in your university site where you may procure this paper as per terms of the assignment.

ON LINE JOURNALS

It is not possible to access each and every journal in any library and researcher has to content with the available journals. However, in recent years many journals are being published "on line" ; this means that any person may access any article of these journals through internet and may check what is going on in scientific world. There are many good on line journals on Parasitology like "Journal of Parasitology Research " or Annals of Medicine and Parasitology while PLOS PATHOLOGY also publishes related papers.. They are good to access even full research paper, but are costly for submitting research papers as most of the journals require a 'processing fee', may be 500$ for an article which is difficult to manage by the Indian institutes.

Number of these on-line journals are swelling every day and have emerged from India too. TROPICAL PARASITOLOGY, run by Academy of Tropical Parasitology, Pondicherry, India is an open access online journal which is also affordable for submitting research papers (www.tropicalparasitology.co.in). These may prove a boon to the institutes which have still not linked with any other national network system of retrieving scientific literature. Though there are only limited online journals on parasitology, yet there are other online journals i.e. international journal of livestock production research,which publishes articles related to parasitology. You may search large number of related online journals by logging www.scopemed.org/ e.g. international journal of livestock research (from India).

SCIENTIFIC SEARCH ENGINES

In recent years many scientific sites on internet have emerged which furnish facility to search scientific literature on their sites. These sources are good to search the literature but it must be clarified that these sites do not explore all the pertinent literature. Rather, they contain only important references, mostly whose reprints may be sold at a price hence searching these sites does not guarantee compiling all the relevant references; moreover, it may prove a costly affair to procure a research reprint from these sites on individual basis. For obvious reasons,, there are few references on these sites of Indian authors.

PUBMED is the site which is free for use and provide screening of large number of journals which is not true for other sites. Here we are mentioning some important websites where scientific work may be screened. These search engines are Google science, Google scholar, yahoo scholar, Wikipedia, Science direct, Scopus, springerlike, CABI, AGROVA, AGRICOLA, EBSCO. Some of these sites may be logged by any individual while many sites like science direct (www.sciencedirect. org) or CABI (www.cabdirect.org) require username and password and are made mainly for research institutes on an annual subscription basis. Obviously, these sites are not specific to Parasitology but deal almost all the subjects of health sciences and many other subjects. Searching of literature now a days has been made easy in every institute as they have been member of one or other knowledge system ; the requirement of journals are being fulfilled by these searching engines.

GENERAL SEARCH ENGINES

Beside above specific scientific search engines, two most popular search engines are Yahoo and Google where you may get a number of information and these may be used for different purposes. Associated with these search engines, Wikipedia is a new form of Encyclopedia which is ready to provide you basic information on almost all topics though some need authentication (many references are taken from this site-see notes).

It may be concluded that in present era, there is no dearth of information and knowledge and internet has made its assessment most easy from any corner of world. Now the challenge is to create new knowledge at fast speed that is capable of changing human life.

ABBREVIATIONS OF SCIENTIFIC JOURNALS

This is an important part of research publication as some universities may require to provide not full name of the journals but abbreviations of the journals while writing them under title references. This is also true for some scientific journals. You are not allowed to change abbreviations or to make them at your will; rather these are taken from the journal itself. But it is not possible to access each and every journal for recording its abbreviated name hence there are some sources which facilitate to observe abbreviations of the journals. Moreover, these sources are not providing only abbreviations of the journals but all the details regarding that journal - year of start, address, change of name of journal, editor, years when not published. Following are the sources to check abbreviations of scientific journals :

1. World list of scientific periodicals, 4th Edition, Buttersworth, London 1963-1965. This volume records all the scientific journals published between 1900 to 1960.
2. World list of scientific periodicals. Buttersworth, London. This volume records new periodicals between 1969-1978.

As you will appreciate that these are the old publications hence do not provide details of periodicals, published afterwards. The University Libraries, Maryland,USA in July 2009 publication (www.lib.umd.edu/guides/scijrnl.htm/) has provided details of the journals in two categories-first general and other is subject specific. Gerry Mckiernan of Iowa state university library has also attempted to compile abbreviations of scientific journals (www.abbreviations.com/jas.asp).

The abstracting journals also provide this facility of showing (generally annually) all the journals which are abstracted in their journals along with their abbreviations. The National Library of Medicine, USA is listing all the journals, with their abbreviations, ISSN numbers, etc which are abstracted in MEDLINE. The details of the journals may be obtained by logging www.ncbi.nlm.nih.gov/nlmcatalog.

This facility has also been provided by CABI, UK and AGORA for Agriculture and Veterinary journals, which are abstracted in their abstracting journals.

In summary, if the scientific journal or university is requiring abbreviated form of the journals, you have to provide correct abbreviations which may be extracted from different sources.

THESIS, RESEARCH REPORTS ETC

At times, you may be given research problem in your post graduate program which is an extension of your previous research scholar. It is a most common practice in our universities that a faculty member is having expertise on one subject and he extends different aspects of these subjects through his post graduate students. Interestingly, the students also join this faculty member after ascertaining his reputation on the particular subject and wish to carry over another aspect of the problem. Thus, my post graduate students have worked on schistosomiasis - it may be immunodiagnosis or comparison of different methods of parasitological diagnosis, porcine or caprine schistosomiasis etc. Our Professor and Head of Pathology Dr JL Vegad was a renowned scientist who has worked on inflammation and all his post graduate students undertook research work on inflammation.

Such arrangement is beneficial to both the sides. The scientist has a vast knowledge of the subject and the department has many related facilities except a few which new research problem will require. Again, with the spread of American education system, Ph.D is not considered a last word in the knowledge but beginning of research career of a student - it is imparted by training the students to solve a new problem using a methodology and the way to defend his thesis or his way of solving the problem before august gathering.

If you are extending the research work of previous research scholars, the theses of previous students are much helpful in all sense; it is helpful in retrieving the literature, understanding methodology and the arguments put forward by the previous workers. Therefore, consulting these theses is important and they are available in the department as well as in the college library. Earlier, the students were asked to submit only hard copies of the theses but now every student is asked to submit the thesis in soft copy form also which is mostly a CD ; this is to facilitate maintenance of thesis copy in the library and easiness in consulting the thesis.

The same may be true if the scientist has worked and submitted a Research project report on identical topic to the funding agency. However, generally the research reports do not carry detailed review of previous work but it is equally useful in understanding materials and methods, results and discussion put forward by the senior faculty member.

It is not necessary that the related thesis may be only from your department; scientists of other universities might also guiding students on same or related topics. In that eventuality it is better to procure a copy of the thesis to understand the work done by the student. Sometimes the thesis copy may not be available, then you may procure summary of the thesis from the student. Haryana Agriculture University has taken the job to publish a biannual journal "Thesis Abstracts " which is providing details of guide, student, title of thesis, year of thesis award and summary of the work ; such information is much useful to get further details. Even the subject specific journals i.e. Journal of Veterinary Parasitology are now publishing thesis summary for the benefit of the students.

However, there is a word of caution while consulting theses of other students.You are allowed to study the thesis but do not try to copy any part of the thesis. At times, it is observed that the student has taken paragraphs together under review of literature or discussion from other thesis. This is not only objectionable but may put you and your guide in great trouble as such acts come under plagiarism and there are instances where whole career of a prosperous scientist has ruined over a small mistake. Therefore, it is advisable not to involve such practices while starting your research career.

DATA ON RESEARCH IN PROGRESS

The Agricultural Research Information Centre (ARIC) of ICAR was created in 1967 and possesses the data base of agricultural research projects (including Veterinary) being conducted under ICAR research institutes, adhoc research projects, data base on all India coordinated research projects are maintained by ARIC of ICAR and may be retrieved by approaching the information systems officer of ARIC.

CROSS REFERENCES

When you are reading a research paper, review article etc you will come across other references which the author has cited to substantiate his results. These references are cited in the text, either with numbers or with author's name and year and all the details of the research papers are appended in the last part of the research article under references. These are cross references, which one has come across by reading an article and important references should be noted down for searching further details of the article.

CONTENTS/INDEX OF RESEARCH JOURNALS

Though procuring a copy of research paper may require payment of certain fee, almost all important journals are having their websites which have the facility to reveal their previous volumes with contents or index of each volume (even some journals provide abstract of the papers). This provides you the facility to check these journals for important research papers, published therein, along with their abstracts.

CURRENT CONTENTS

This is an old, prestigious weekly publication which was informing table of contents of research journals so that researcher may observe the title of research papers in a shorter time and may ask to provide full research paper, required for their research work. However, subscription of this journal was heavy and there are only a few institutes which are subscribing current contents.

REPRINTS OF RESEARCH PAPERS

Abstracting journals help you to search the literature and to read abstract of the research paper. The abstracts provide important information related to the research paper; it informs briefly on materials and methods,results and importance of the result in present contest. But abstracts of any paper fail to provide detailed information of any aspect of the research carried out by the scientist. Therefore, to know the details on materials and methods, results and discussion, study of whole research paper is necessary. The research paper may be searched in the library following details of the journal, year of publication, volume and page number.

At times, the particular journal is not available in your library but the paper is important for your research work. In such cases, you may make a reprint request. This request may be directed to the author of the research paper, particularly if he has published extensively on the subject hence you may request to send other relevant research papers too ; all this you will get free of cost with the kind courtesy of the scientist. As mentioned above, ICAR is providing the facility of sending reprints, free of cost, but only of Indian journals.

The other source of getting the reprints is our National Libraries where a request is made by enclosing demand draft for the reprint. There are four important national sources to whom you may depend on your reprint request :

1. National Library, Indian Veterinary Research Institute, Izatnagar, Bareilly.
2. Indian Council of Medical Research, Ansari Nagar, New Delhi.
3. INSDOC, New Delhi.
4. Agricultural Research Information Centre (ARIC), ICAR, New Delhi.

Generally, these organizations take 1-2 months in dispatching the reprints. Alternatively, you may ask your colleague, studying in same institute to dispatch the reprint.

If you are searching the literature through internet, there are various sites which cite references (see above). They provide abstracts of the papers while only a few papers (particularly short notes) may be seen with full text whereas others are at a cost. For Indian scientists, the price will appear mostly exorbitant as each reprint will cost about 30 $ though these price are reduced in case your institute is subscribing the site of the organization.

RESEARCH IN POST GRADUATE COLLEGES

There are many degree and PG colleges in every state, at district level,that are running mainly humanities though some colleges are also imparting courses in biological sciences and even Ph.D is run by some institutes. These colleges have only limited library and laboratory facilities hence I will not advise any brilliant student to join these colleges for doing MSc or Ph.D in biological sciences. They may not have proper laboratory facility and periodicals that are present in the universities. Even if you try to acquaint the latest scientific information through internet it will only be partial as these search engines, free for all, are for general public and not for scientific community.

These colleges may be joined only when you are acquiring the degrees not for pursuing research as a career. Later if you plan for research, you have to work hard to break the barriers and being absorbed in some other prestigious institute.

15

Conducting Research

Review of literature or consulting of literature is just beginning of a research project. The main intellectual work is to study the previous work meticulously and to draw conclusions regarding existing problems. This leads us to another intelligent part of the research i.e. planning of the research work. If you analyze what you are planning, it is only about materials to be employed and methodology of the research work (with planning of the work)- only these three parameters are under your control as results will be the outcome of these three facets while discussion will be how you consider your results in presence of previous research work. In fact, this statement of mine is more align to the sermon of "Shrimad Bhagwat Geeta" (5) where it is stated that man has control only on his actions (planning and conducting research) hence fulfill this duty without caring for the results-which in turn will depend on your analyzing the previous facts and how you have planned to solve the problem. Therefore, pay full attention on these three facts of the research.

STUDYING PREVIOUS RESEARCH WORK

After searching and recording the literature, it's necessary to assort the literature in different categories and study them deeply as this exercise is essential for planning your research program for solving a particular problem. This simply means you have to analyze the previous hypotheses and note the pit falls in those suggestions.

If we go back in nineteenth century, scientists were aware about association of malaria with 'foul air' and since mosquitoes were in abundance in these places, it was suspected that mosquito might be transmitting this disease to human beings. If this hypothesis was true even there were many other related questions which are to be solved e.g. Which mosquito stage (egg, larva, pupa, adult) is responsible for malaria; how mosquito get the organism from infected man and how it transmits to a non-infected person ? Whether it was water polluted with mosquitoes,responsible for malaria ? Whether all types of mosquitoes are transmitting the organism or only a specific type ? How it is transmitting the infection- whether mosquito excreted the organism in the water and man get infection by drinking the water or man get infection by accidentally ingesting the infected mosquito ? Or there was any other way ? There were various hypotheses existing at that time. After some experiments, it was suspected that mosquito bite is responsible for malaria; and with this idea, Ross began collecting mosquitoes after bite of a malaria patient and tried to infect another person by feeding these mosquitoes but failed etc Slowly, he realized that there are only certain types of mosquitoes which are spreading malaria. Meantime, there was a report from his mentor Dr Patrick Manson from Britain about finding filarial larvae in a mosquito;however, the basic question remained how and when this mosquito is capable of infecting another person. He conducted several experiments and ultimately was able to proof that mosquito gets infected by its bite to a malaria patient and again a mosquito bite is able to transmit the infection to a normal host, provided the mosquito has developed infective malarial stages. As every one of us are aware that Sir Ronald Ross was awarded Noble prize in 1902 in Medicine for this path breaking research. This work clearly suggested how Ross remained aware about contemporary activities and also his original thinking which led to finding the solution. Thus, both facts are important in big research- your being well aware about the problem, how the problem is being solved by others, their failures, and then your original thinking to tackle the problem. Therefore, you have to study each and every aspect of previous work done by your contemporary scientists.

MATERIALS AND METHODS

This is the main part of our problem where we have to analyze methodology followed by previous workers. This includes two facts. The first is the material used by the scientists. If he has used laboratory animals in his research work, what species of animals were these, whether interbred or random, their age and sex and how each of these factors might have influenced the results. If the work is related to the field all the details of the host should be analyzed,and to note if any one of them might affect the outcome.If it is related to search of a snail for a fluke infection, how the scientists identified snails, different snail species encountered; chances of existing strains,their age or size, locations like stagnant water bodies, temporary or permanent water bodies, water levels from where fauna and flora were collected etc.

Another important factor, in methodology, is the technique employed by the previous scientists. It should be minutely studied and investigate the lacunae in the techniques used by the previous scientists. In first case, study the techniques used in the experiment and the factors which might have affected the output. For instance, recovery of schistosomes from infected animals by cutting and soaking tissues in saline, or teasing blood vessels or using perfusion technique -all will greatly influence percentage fluke recovery hence results of the experiments. In identifying snail intermediate host, what techniques were used to check positivity of snails for trematode infection- exposure to light (will fail to shed nocturnal cercariae), teasing the snails, examining under a compound or stereoscopic microscope etc.

In analyzing literature on immunity against schistosomes, I observed that the Indian scientists have neither used anti-coagulants nor any perfusion technique for recovering schistosomes from experimentally infected animals though the same have been used abroad. Importance of anti-coagulants in recovery of schistosomes was established by us by finding the schistosomes in blood-clots (2). So was the case, when a perfusion technique was used for recovering blood flukes from the experimental animals.As quantitative recovery of schistosomes is essential in immunity or drug trial research, use or no use of these parameters are important though may not be in determining host susceptibility for a given parasitic species. Should we, therefore, discard all the immunity research work carried out without anticoagulant or perfusion technique ? Not really.

This is because scientific research has a procedure to negate common factors in hindering significant outcomes from these experiments else all our old experiments would have been condemned. And this is to keep a control of the experiment where all factors except one is made identical thereby negating effect of old technique (which is used in identical ways in such experiments) used in the previous experiments. In above case of immunity experiments, two groups of the animals are kept, one infecting without immunological dose and other infecting with immunological dose; method of recovering schistosomes from both groups is made in identical way thereby excluding the influence of recovery methodology reflecting the main effect because of studied factor (development of the immunity). The comparing results, using old and new techniques, also hints strong immunity development in schistosome infection by an earlier infection. However, there is no denial that a bad technique would have not been able to identify a weak immunological response hence advocating importance of an effective technique in checking any fact.

The above facts point out one important development in science i.e. the technique of investigation is improving day by day and so our results are also refined as new experiments are conducted with the use of new techniques. In fact, advancement in science has always been associated with development of new methods for dealing an old problem. Earlier, in taxonomic work, parasite species were created when ever any minute morphological difference (e.g. number of spines, spicule size, size of worms) was noted in the parasites, or parasites were recovered from a different host species and such work was published in national journals; but now its not possible to publish such work,unless and until such observations are supported by other techniques- PCR studies or biological or biochemical differences between the two specimens.

Another example that may be cited is infecting large animals with schistosome cercariae. In olden days, lambs and pigs were infected with schistosome cercariae, 10000 to 20,000 per animal, either per os or keeping the animal in plastic tub,filled with infected water (16). Because of this technique, such a high number of the cercariae per animal did not cause death of any animal or much pathology suggesting as if schistosomes are less or not so pathogenic. On the other hand, when piglets and goat kids were infected by a new method using polythene bag and tail or ear-pinna of the animal, a much smaller dose of 1500-4000 cercariae in each animal, was sufficient to cause death of the animals or more sever pathology in them (60). Thus,this new technique of infecting large animals was not only simple but more accurate where almost all schistosome cercariae were able to penetrate skin of the host. However, when you are analyzing results of the above research papers, remember that the difference in results is due to difference in infecting the animals and it will be a wrong conclusion to draw that even 20000 schistosome cercariae may be harmless to these host species. These conclusions can be drawn only when you study the previous literature more carefully.

RESULTS OR OBSERVATIONS

After scrutinizing materials and methods, the results of previous workers should also be critically evaluated. You have to evaluate whether the scientist has used correct method and materials for arriving that conclusion (results) or there may be some gaps in methodology hence results. For example, Sinha and Srivastava (56) conducted host susceptibility of different animals (albino mouse, rabbit, rat, pig etc) against *Schistosoma incognitum*. They observed development of adult flukes in albino mouse with presence of its eggs in liver and intestine but not in the feces. Therefore, they concluded " in white mouse, though the worms attain maturity, their size was comparatively much smaller and their eggs could not break through the wall of its intestine to be voided with the feces". The subsequent workers (including me) supported their view and white mouse was considered an unsuitable host in which the infection remained in its latent form; for this reason, patency was not a criterion for judging immunity against *S. incognitum*.

When we took up again the problem in 1984, we minutely scrutinized the facts and it appeared most unlikely that albino mouse is not excreting *S.incognitum* eggs in its feces. If yes, what may be the reasons for their failure and two most probable reasons emerged. First, as excretion of eggs in pigs (natural host) starts from 30th day of the infection, they examined feces from 30th day and may have extended it up-to a week or so and not beyond that period. Secondly, hatching and direct smear methods are relied in the pigs the same were also relied in white mouse without following any other specific coprological method and when they failed to detect the eggs in the feces,they hypothesized the reasons for the same (ironically, the workers have not provided experimental details of days and methods of fecal examination in their research paper).

In our experiment, we infected 12 white mice,each with 500-600 *Schistosoma incognitum* cercariae and killed 6 mice between 4 to 6 weeks to confirm development of the flukes and egg concentration in liver and intestine. The feces of all mice, on 30 day post infection and afterwards, were pooled and examined by two methods-hatching method and a more sensitive acid-ether method. Though hatching method remained negative for any miracidia, the acid ether method detected *S.incognitum* eggs in the feces on 49th day post infection and thereafter individual fecal material was checked; this confirmed passing of the eggs by each and every infected mouse (2,57) though hatching method could not detect miracidia in the fecal material (however, intestinal washings during necropsy showed large number of the miracidia). Thus a deep study in the case, has resulted in new findings and subsequent workers have confirmed our findings.

I may cite another example how intermediate host of *Stephanofilaria assamensis* was discovered (85) by Dr SC Dutt in early sixties (a part of research work for which he jointly was awarded national Rafi Ahmed Kidwai award of ICAR). Prior to his research, life cycle of any *Stephanofilaria* species, prevalent in any country, was unknown. For other filarial worms,a mosquito species has been incriminated and scientists were considering same for *Stephanofilaria*; however, Dr Dutt studied the literature critically and noticed that none has found mosquito anywhere as intermediate host for *Stephanofilaria*. Therefore, he kept his option open that there might be another arthropod or fly responsible for transmitting this filariid. In his experiments, he inspected many hump-sore cases (caused by *S.assamensis*) and collected flies which were visiting hump-sore lesions. He examined them in his laboratory and surprised to note that there was no mosquito but mainly *Stomoxys* and *Musca* flies. Though *Stomoxys* is a known blood sucker (earlier incriminated for transmitting blood protozoan trypanosomes in bovines), but till then *Musca* was never incriminated as blood sucker and was considered a lapper–depending on semi liquid food. But since these *Musca* species were found on the lesions, he examined all these flies with other flies, after making preliminary identification, for presence of any larval stage of *Stephanofilaria assamensis*. Ultimately, he found intermediate host for *S.assamensis* and surprisingly it happened to be a *Musca* species- *Musca conducense*. But he has to provide explanation for his findings that how *M.conducense* can act as intermediate host for *S.assamensis*. The explanation was that the microfilariae of *S.assamensis* are present in the epidermal part of skin and not in blood circulation (unlike to other filariids) and this species of *Musca* has a proboscis,sufficiently strong enough to tease this part of skin and imbibe oozing blood along with the microfilariae (which converts into infective third stage larvae in *Musca conducense*). Later work by other scientists have confirmed his findings.

There are many examples in the scientific research where a critical study of previous results and devising new experimental designs have resulted in new findings. The only fact is how minutely you have studied the research papers, published by previous workers and have designed your research plan including new ideas for conducting research.

DISCUSSION

Studying the discussion part of a research paper is also important as it provides you justifications for the present results and explanation for the difference in results between the scientists. When a scientist is differing in the opinion with other, it is prudent to study critically the materials and methods and results mentioned by both and what may be the reasons for leading this difference of results. It is again not necessary that every scientist has conducted his work seriously and you may exclude such findings in your consideration. For example, we were trying ELISA in natural schistosomiasis cases in domestic animals and have never attained above to 90% specificity and sensitivity. However, in one reference the scientist observed all cattle positive by ELISA which were excreting or not excreting eggs but the test was negative in cattle which were never exposed to schistosome infection. Similarly, we have observed one reference on human schistosomiasis from India where ELISA was hundred percent positive in egg excreting cases and negative in non-excreting cases. In fact, such cases are excluded and considered by many as providing erroneous results since they have not provided any convincing explanation why their results differ from those of others.By reading the discussion, you may find out the lacunae in their point of view and this may form basis of your future research work.

FRAMING RESEARCH PLAN

When you have critically studied the work of previous scientists, it should be clear how they have attempted to solve the problem and what are the gaps in their research work. Some idea will flash in your mind and you will be able to plan your research work. As mentioned above, you are planning only three parameters of your research- one is materials and other is methodology and research planning as results and discussion will be outcome of these three.

MATERIALS AND METHODS OF RESEARCH WORK

As mentioned, the plan is related only on two facts. The first is selection of our materials and other is methodology of the research work- we cannot confirm outcome of our experiments though we may predict them but at times we receive surprising results; therefore, all our research planning, though employing all our intelligence and diligence, is related on what materials we are selecting for our research work and how we are examining them.

MATERIALS

The material will depend on type of your research work- it may be animals, chemicals, biochemical kits, biological molecules, parasitic material, a pond, a grazing place etc Its better to attempt to collect your chemicals prior starting your experiment as this may take time. Rest assure that your laboratory is possessing all the materials,required for the experiment or place the orders in time so that it may not delay your experiment. The materials may also be a special equipment or machine which is needed for processing the materials.

METHODS

The other important fact in the research is what methodology you will use in your experiment. Your methodology should be standard so that questions should not be raised over its efficacy; if the methodology has been developed in your laboratory, you have to prove that it has been standardized against a well known technique.

Your laboratory may not have a particular equipment or special biological marker which is must for your research but not possible to acquire the same due to various reasons. In such eventuality, it is advisable to contact the institute where such facilities are existing and to request for conducting part of the experiment in that laboratory.

Beside these materials and methods, you have to plan your research where you have to take care of certain factors. Some of these are discussed herewith.

PARAMETERS TAKEN CARE OF

In any biological research, please remember two important facts; the first is to take sufficient number of observations (see below) so that it helps you in drawing any conclusion. This sufficient number differs according to the subjects you have taken in your experiment- number of laboratory animals will always be more than number of domestic animals, you have selected in the experiment. If you are finding an intermediate host for a parasite, the examination should be in thousands of the vectors prior reaching to any conclusion.

The other important fact is repetition of the experiment as single observation is not accepted in scientific domain. This is carried out by carrying a number of similar experiments simultaneously that saves time in carrying the experiments. Thus if you are trying to check efficacy of a drug against a parasite in experimentally infected animals, a minimum of 6 to 8 laboratory animals are treated with the drug and identically work is repeated in six or more animals; again sufficient number of animals are kept as untreated controls; though use of animals have come under new rules and regulations with changing conditions and without approval of appropriate committees, you are not allowed to experiment on animals (see below).

Constant factors

When ever you are planning any experiment, it will be advantageous to know which factors may affect your results and be cautious about them while planning your results. For example, it is well known that fluke infections spread with the help of fresh water snails, therefore, certainly there will be difference of prevalence between grazing animals and stall fed animals. Therefore, it is important to keep these two groups of animals separate and you may calculate your results in each group to show effect of grazing on fluke infection. When you are searching cercarial dermatitis in a village population, you have to depend on those persons who are using water bodies for different works; If it is known that particular parasitic infection is occurring in juveniles then you should plan your experiment in such a way so that this factor becomes prominent-either you undertake work in juveniles and divide the group in different categories like male, female, race, health etc and compare the infection rate between these categories. Or you take work in higher age group to claim up to what age the infection persists in that subject species.

If you are planning to know whether an infection is transmitted from mother to the offspring through placenta or breast milk, you have to examine the neonates and offspring prior and after breast feeding; this work is strengthened by examining mother's milk for infective stages of the parasite.

Thus, there are certain factors which are known to influence your results and you have to consider such factor while planning your experiments.

Number of cases

This is an important consideration having power to alter your results. If you are planning to know difference in prevalence rate of any parasitic infection between male and female, you have to select equal number of the hosts of both sexes in sufficient numbers which at times becomes difficult while studying in our domestic animals. Again both male and female should live under identical conditions. For instance, if you are collecting data of hookworm infection in man and woman (to access if sex hormones play any role in such cases) both should be using fields for defecation and if this is not so, the variation in infection rate, if detected, cannot be assigned to sex differences or sex hormones.

The number of cases while calculating prevalence rate or alike parameter in a given population should be sufficient so that it may represent the whole population. If you are calculating parasitic infection in a village population, at-least one fourth population or a minimum of 200 should be involved in your research plan.

This number of cases may increase provided the infection rate is reported to be low,as observed in some areas. For instance, we are aware that only 1-2% snails turn positive for schistosome cercariae in a given locality. And if you are searching intermediate host for a schistosome species, it is not acceptable to declare your results by examining only 50 to 100 snails of that species. For a just inquiry, the number of snails of a particular species should be between 500-1000 snails than only your results may be regarded justified.

In experimental parasitology where you are using experimental animals for verifying certain facts, you have to prepare at-least two groups of animals- both are infected under identical conditions and in one group treatment is provided while other group is kept as control -without treatment. Ideally, each group should have atleast six animals but if you are using domestic animals like pigs, sheep,goat, calve, the control animals may be reduced to four or so. At times, looking to stringencies, control group is totally abolished, particularly when we are dealing large animals- though keeping of a few animals as control is always advocated. In former cases, prior treatment data are collected and they are compared with post treatment data. For instance, we are planning to study hematological changes due to a particular parasitic infection. Sufficient data are collected from the animals prior infecting them with the parasite and then data are generated after infecting the animal. If there is significant reduction in hemoglobin concentration after infecting the animal, it is assigned due to the particular parasitic infection.

In summary, number of animals/parameters should be sufficient so that statistics may be employed on your data to draw conclusions by applying statistical analysis else your research work may face problem,even in publication in a national research journal.

Random selection

This means you have to select randomly your subjects other wise your research will be biased. For instance, if you are planning to observe *cryptosporidium* infection in buffalo calves up-to one month of age, you have to select buffalo calves up-to one month age randomly irrespective of their sex or health or breed and your experiment should include calves from all these categories; otherwise, you may plan your experiment to investigate *cryptosporidium* infection in male calves or murrah calves of one month age etc.

If you are checking snails for certain fluke infections, you have to collect the snails randomly from the habitats from upper water level as well as from bottom. Again it is important whether you are collecting the snails from the locations where final host is visiting the habitats or your collection is from the place where approach of final host is difficult.

TWO TYPES OF RESEARCH WORK

The research work which you plan may be divided into two categories :

1. Field problems
2. Experimental problems.

At times, these works may be intermixed where you have to carry over your work under both conditions.

Field problems

We have already emphasized earlier (Chapter 7), the importance of critical observations in the nature which have been responsible for land mark discoveries. This work may be required for testing a new drug or a new vaccine or a new diagnostic method in different host species under different ecological conditions prior of approving the same for mass use.

Alternatively, we have to study nature, when we plan to solve a problem existing in nature. For instance, we are aware that fluke infection cannot exist in a geographical area in absence of a snail host hence finding a snail host species to a newly discovered fluke species will be our field problem which will require studying contact behavior, habitats inhabiting snails, chances and reasons for visiting final host to these sites,collecting all types of snails, their classification,searching cercariae or metacercariae in snails, identifying them and finally developing in intermediate and final host to recover identical fluke specimens. Thus you will observe that whole problem was started in studying and collecting material from nature and culminated to experimental work to confirm certain findings.

Likewise, there are many other problems which require field work. Studying of seasonal occurrence of a parasitic infection in different host species will require field work; parasitic prevalence as per age, sex, breed of the host will also require field survey; parasitological work involving slaughter house material will be partly a field work.

An important aspect of parasitic diseases in any host species is to study clinical symptoms and sequences of the disease; all this work has been carried out by the clinicians observing the patients day and night and this is foundation of parasitic diseases both in humans and animals. There are many details of symptoms and lesions of every parasitic disease documented in our text books. At times, such work has been associated with experimental work to confirm some parameters of the disease.

Its not simple to get correct results from field work with all chances of criticizing the work if minor details are not followed. For instance, if you have recorded prevalence of an infection month wise, without taking metrological data of that area,your work will have little value; prevalence of a fluke infection in domestic animals without taking husbandry practices (e.g. grazing vs non-grazing) will provide confusing results; difference in prevalence rate between male and female may not solely due to sex or hormonal variation; slaughter house studies can not be considered representing that areas' prevalence since slaughtered animals (age, sex) are belonging to different areas and without knowing source of animals, coming in slaughter house, you can not assign the infection to the area where slaughter house is located.

Field work is also essential for finally launching a product in the market. These field trials are conducted under strict controlled conditions and by the experts, first on a smaller scale and later at larger scale. There is general practice to use a standard drug and a placebo (material without drug) for checking new drug efficacy; the efficacy of diagnostic kit is accessed in known positive and negative cases, judged so earlier by an confirmative diagnostic test; whenever vaccine trials are tested, the subjects are divided into two groups- one vaccinated with the new vaccine and other without vaccination and development of the infection in the two groups is accessed after stipulated time. Its not necessary that our laboratory findings will always simulate under field conditions and many times opposite is observed.

Experimental problem

This is a very strong branch of Parasitology which has emerged in nineteenth century with the perception that we can find many solutions by conducting different experiments under controlled conditions. For instance, if we wish to understand life cycle of a parasite in final host, the experimental parasitology is helpful in studying the detailed parasite development with the organs,affected by the parasite and one can study the time taken by the infective stage to different stages reaching to adult stage and excreting eggs in the excreta.Simultaneously, symptoms and lesions of the infection may be closely co-related with the developmental stages of the parasite and specific time taken for developing each stage, which is not possible in natural infections. This is the experimental parasitology which co-related cough and lung pathology due to migration of larvae of hookworms in the lungs, the infection which otherwise is known to cause dysentery and anemia in the infected person. In fact, experimental parasitology has helped in understanding life cycle of many parasites, many biochemical changes occurring at particular stage, how and what stages are pathogenic to the hosts- both intermediate and final. This has also helped to develop laboratory models where many studies, particularly immunological, pharmacological,pathological, are made in important parasitic infections.

One more reason for developing experimental parasitology is strong centers of Parasitology in European and American countries where many tropical parasitic infections are not existing hence many Parasitologists of these institutes have diverted their attention on basic parasitological problems which require strong laboratory facilities. Thus development of vaccine against Malaria or schistosome is being carried out in these institutes though both the diseases are the problems of African and Asian countries.

With the beginning of twenty first century, some new restrictions have been imposed while conducting biological research, and you should be aware about these regulations else you may face problems and your research paper will not be published by any peer reviewed journals.

ETHICAL COMMITTEE

If your research plan involve human beings or domestic animals, you have to put your whole research plan before ethical committee of your institute which will scrutinize its validity. If you are using any invasive technique like drawing blood from the person, or giving injection to any person, you have to provide all the details and also why such program is necessary in your research work. You have to explain all the precautions, taken so that any accidental transmission of any pathogen would not be possible. At times, you may be denied to use your procedures in the prescribed population. Generally, there is little restriction if your work is merely collecting human excreta for some survey work.

PREVENTION OF ANIMAL CRUELTY ACT

Earlier the experiments on laboratory or any type of animal were conducted without looking after cruelty factor on the animals. If you review the old research papers,you will notice use of a large number of animals for deciding merely one factor i.e. host susceptibility for a given parasitic species. There was no pooling of the experiments which was possible; thus if we are infecting 10 goats with amphistomes/fasciola/ schistosome, the same animals may be used to study nutritional requirements, biochemical studies, applying different diagnosis methods,chemotherapeutic trials and if animals are to be sacrificed, the event may be used both by Pathologists and Parasitologists- obviously this was not happening previously. Now, prevention of animal cruelty rules have been notified on 26th March 2001 (www.moef.nic.in/legis/ awbi/awbi18.html) and you have to present your research plan before a college committee to justify use of animals in your experiment and that you have taken all the precautions to prevent cruelty against animals; its approval is important as without such certificates, the research journals are also rejecting the research papers. Your organization might not be having an animal house to supply the animals for your experiment; in that case you have to order for animals along with certificate of approval of your experimental plan. Obviously this will take some time hence you have to plan purchase of the animals, in advance to avoid the delay.

PATENT LAWS

This is about making patents of your research findings. This may include any procedure or technology. Thus, if you have developed any new technology which has some commercial benefit, you may proceed to patent it under patent laws of government of India (86). However, the same law restricts you to use any procedure without taking permission to use a patented method. Every university is supposed to have a Patent committee to look after these aspects and helping to apply for patenting your research findings.

NO REPEAT WORK

Time is changing and its futile to conduct the work which is a repeat work as many peer reviewed journals do not publish them or at the most will publish it as a short note. Again do not select the research work which has little consequences. For instance, determining prevalence rate of a parasitic infection in a host species in a given geographical area has little consequence as there are many reports on this subject and there is nothing new which may come out except some change in infection rate. If at all, you are interested to repeat such work, always plan to include some new parameter which is important scientifically with this old work. For instance, if you are calculating prevalence rate in your area in domestic animals, you may co-relate it with some biochemical changes,occurring in each host species, or economic parameters like milk production or body weight loss provided these parameters have not been taken up by previous workers.

It is always encouraging to find intermediate host for a parasite whose intermediate host is still unknown; for instance, though *Bivitellobilharzia nairi* is widely prevalent in the elephants,its intermediate host is still unknown. You may take up the problem where a controversy is existing as is in the case of *Ferrissia tenuis* - this is incriminated experimentally in the life of *Schistosoma haematobium, (S. gimvicum)* but has never been seen shedding schistosome cercariae in any area,though the snail is widely prevalent through out India. You may take up the research work where you have to find intermediate host in India though it is known in other country e.g. *Orientobilharzia harnisutai* is suspected to exist in India and it has been demonstrated in buffaloes, only once, and its snail host is still a mystery in India though *Lymenae gibsoni* is incriminated in Thailand (87) (the original source of the parasite).

TEAM RESEARCH WORK

When ever you are dealing a big problem, it may become a life time devotion to that problem. Many such problems are difficult to solve by one stream of scientists and it is always advantageous to associate with scientists who are expert in related fields. If you are trying to develop a vaccine against a parasitic infection, you may have pure immunologists in your research team; if you are working to develop a resistant sheep strain against *Haemonchus*, it will be better to have a pure genetist or animal breeder in your research team and so is the case with other fields.

It is neither advocated nor possible to change field of research frequently as this does not lead you anywhere. The present trend is to stick to one field of work and devote whole time to find solutions of different problems,arising one by one.

THESIS PROBLEM

It should be clearly understood by a post-graduate student, that thesis work is a training to a student to understand how he can solve a given problem, so that he can take up the problem, later, independently and come out with solid results. This is the academic research to be conducted for acquiring either Master's or Doctor's degree in Parasitology. Therefore, the thesis problem will mainly depend on the facilities available in the department as well as area of research where guide is having experience. Nevertheless, there is a choice to the student for taking admission in the college where research of his interest is being carried out. It is better to make correspondence with the Head of the department showing your interests and desire of seeking admission in the department. Though foreign universities have the provision of allotting the guide of student's choice, this practice is still not followed in the Indian universities and there are many factors which determine allotment of the students to each staff of the department. Again it is advantageous to be aware about the basic atmosphere in the department as in some departments the staff members keep a long distance between themselves with little hope of help from other faculty members. The students should be cautious in selecting their post-graduate guides and always select those who are doing research work and publishing them; it may be problematic if your guide has yet not published any research paper as this suggests his least involvement in research activities except that thesis guidance is thrust on him as one of departmental duties. This is more important if the teacher is either associate professor or assistant professor.

We have witnessed the research scholars who could not publish even a single research paper from his thesis due to cold response from his guide. This may be affordable to those who have enrolled for post graduate degree only to get further departmental promotions or getting one or two increments but certainly is detrimental to those who are choosing research as their future career.

It is advantageous to analyze the research problem prior making it as your thesis problem. There are certain points which will help you in this decision :

1. The thesis problem you are selecting should be dealt with in the facilities available in the department or there is provision in the college of taking out side technical help.
2. Your guide has sufficient research experience and has published research papers on that topic.
3. The thesis problem should enable you to learn as many new techniques as possible and to handle modern equipments as well since you have to deal them in your future research program.
4. Your research problem should not be old one where sufficient work has been published. As attempting such problem will neither help you in learning new techniques nor publishing will be smooth.
5. It will be advantageous to divide your research project in three-four parts so that each may be published as a separate research paper.
6. There are problems which require one year study like seasonal variations etc; in such cases, its better to start your research work earlier so that you may finish it within time.
7. Select a problem which is completed not before six months time otherwise your experiments may be of short duration and will be criticized for small work.

It may also happen that the department is running a big research project sponsored by some scientific organization and you are allotted part of this as your thesis problem. Such arrangements are beneficial to the students as they get all the laboratory facilities developed to carry over research project. There are also chances of getting research scholarships in such projects. In my opinion, it is always advantageous to join such departments who are involved in running big research projects.

16

Recording the Data

After planning research problem, the next step is to start real research work. Indeed, it is said that planning of a research work is only 20% of work done while conducting the experiments is 80% to be done and is more laborious, requiring extended time and patience of the scientist. While conducting your research work,it is essential to record each and every observation of your work. These observations will be of two types. One is the subjective observations which are difficult to quantify. For example, if you are recording clinical parameters of a disease,you may record pulse, temperature, respiration in arithmetic form but clinical symptoms are subjective and require a description of the patient rather than filling tabular sheet. In describing morphology of a parasite both parameters are needed ; you will describe morphology of the parasite and will also mention measurements of different parts of the parasite. Likewise, you may like to describe weather conditions or behavior of animals, vectors at certain stage of disease while you have to mention figures in many cases. Thus your observations are clearly divided in two categories - one is subjective observation and other is recording arithmetic values.

It is recommended that you carry a register or a diary which may be divided in two parts- one part will contain your subjective descriptions while other part will show figures or data of your experiment. Always remember to mention date of your observation and also stage of infection, if you are dealing this aspect.

No one recommends use of loose papers for recording the data or descriptions as there are all chances of missing them at one or other time. It was also advocated that its better to keep two copies of your records for safety purpose. Now, with the advent of computer you may use tablets to enter your data at appropriate places but again remember to protect your files by inserting passwords.

For recording the data, you have to prepare tables where rough data are recorded. It should contain details of each and every value, taken by you. If you are infecting birds or animals with some parasitic material, every details of the host like sex, age, weight, breed and of parasitic material like source of collection, dose, rout etc should be recorded with time of observations and form different tables for each parameter. If you are taking body weight of the subjects or doing hematological work, you will record these parameters for stipulated time and for each subject ; also mention time of taking body weight/blood, prior or post meal with or without anticoagulants, measuring tool or weighing machine etc. Your table should include as many columns as possible so that no data should be left. Many times, we remain unaware about some important fact but realize it afterwards and will look whether our data support our contention or not.

Obviously, these are rough data and tables which are to be used for statistical analysis. Some times, these rough data are considered important in thesis work hence appended by the students so that these data may further be used, if needed. Statistical analysis is important to draw some conclusions from your work

STATISTICAL ANALYSIS

What so ever, you have recorded in your research work, they are termed as raw data and they can not be published as such because of space occupied by them and difficulty to draw any conclusion from these raw data. Therefore, these raw data are compiled in more systematic ways and at-least some basic statistical analysis is essential to carry which will reflect correct interpretation-this may be in form of calculating mean, measures of dispersion, percentage and drawing graphs for depicting the highlights of the results. For instance you may write percentage of infection in a given population.

In present days, comparing results on percentage basis carries little importance and there are chances that some journals may ask for more statistical analysis. However, this presentation in percentage has been a trend settler which was efficient enough to demonstrate importance of scientific work. First time, this simple comparative analysis was used by Mather and Byolston in Boston, USA in 1720s to evaluate a medical procedure and subsiding their criticism of efficacy of variolation (inoculation of material from small pox sac with the help of a Lancet to a healthy person-chapter 2) in checking small pox between naturally contracted and variolated individuals (88). These persons statistically showed that fatality was 15% in naturally contracted disease while only 1 or 2% persons died who were variolated by inoculation method (88). Thus, even a simple statistical analysis was able to convene an important message and helping to some extend in subsiding the controversy.

Statistical analysis is the important step which is able to provide a direction about your results e.g. whether the infection is related with age or sex of the host or your vaccine is able to protect the host or it has failed to protect the individual. These mathematical calculations make sure about your results as these results are the basis of writing any research article.

Beside simple percentage calculations, there are now other statistical tests when you wish to compare two groups of the experiment, and for judging whether your results stand test of significance; how many times there is the probability to get the same results; if same results will be obtained 95 times, the t test is called significant. For instance, if you are comparing infection rate in young and adult persons and your t test turns significant - it suggests that age is playing crucial role in determining infection rate; if the test turns non-significant, it will suggest that it is not age but other factors or chance is responsible for such differences and age is not important in prevalence rate of that particular infection.

An important statistical parameter is calculating standard error which helps in testing whether the difference between observed (control group) and expected frequencies (treated group) arise due to chance or treatment was responsible for the difference. If a difference is less than 3 times the standard error (SE), the difference is supposed to exist by chance but if difference is equal to or more than 3 times SE, chance fails to account for the difference and treatment was responsible for it and we call it significant.

If there are more than two variables, instead of t test, Chi square test is of choice. Another important statistical analysis is ANOVA or analysis of variance and is used when multiple sample cases are involved. For instance,we want to compare hemoglobin concentration in six groups of animals, treated differently; here application of ANOVA will inform whether difference in hemoglobin concentration is because of treatments or attributed to chance (non-significant results).

For a pure Parasitologist, it may be difficult to calculate beyond mean, mode, percentage, although computer soft ware has greatly helped in calculating other statistical parameters. Alternatively, you may consult a preliminary book on statistical analysis (89). It will also be advantageous to take help of the statisticians for applying different statistical tests and their interpretations.

There is a school of thought who have been critical on statistical analysis in biological sciences, particularly in health sciences. According to them, value of life is more important than showing if the available drug can save life only of 30% population which may prove non-significant on statistical ground. Likewise, if a newly developed malaria vaccine can protect only 30-40% children in an endemic area; should we condemn the vaccine only on the ground that it fails to protect large mass of population. Does it mean that we should not save life of 30% persons since they are non-significant constituent of the population. If we study history of a complicated surgical operation, either related to heart or brain, its efficacy or survival rate of operated patient was minimum-many times at non-significant level, but slowly the technique has been improved to the level where failures are non-significant. Certainly, science would have not achieved such efficiency, if the surgical technique has been discarded at its early stage, considering it being a non-significant technique. Therefore, it should be on the discretion of the scientist whether he considers his research findings important or not, though statistical analysis suggests him to discard the path because of non-significant results.

TABLE FORMATIONS

After making statistical analysis from rough data, it is imperative to present important data in table form, for research paper, which is different from the tables which have been used for recording day to day data. These tables are prepared to present only important data mentioning whether results are significant or not. While preparing table, its columns should be fitted to the page of the journal; generally journals ask the tables that cover only half or quarter size of the page. The tables are loaded with optimum values which are readable and acceptable to the readers. It is also kept in mind that only important figures are mentioned in the results and text and table should not repeat the same items. If there are only few values to be presented in the research paper, many research journals ask the scientist to present them in the text and table, in such case, is deleted.

GRAPH FORMATION

Many times, it may be cumbersome or difficult to the reader to check the table for large number of data hence the results are depicted in the graph form so that readers may visualize the changes of a longer duration with little effort than vetting each figure of the table. For instance, if you are studying weekly change in body weight or hemoglobin for a year in a host species, it is of little use to depict all these data for 52 weeks in a table form as the reader has to read all the figures to have a fair idea; rather it will be simple to depict the data in a line or bar graph form along with levels of significance as the same may be interpreted more easily by the readers. Likewise, if your research paper has compared a number of diagnostic tests for efficacy in diagnosing a parasitic infection, it is always better to depict these results in bar diagrams. Fortunately, axel configuration of computers have the facilities to make all these calculations along with the facility of drawing graphs with different configurations and the best may be submitted to the journal for publication.

PHOTOGRAPHS

It is advantageous to take photographs of all places,processes, equipments randomly as we do not know where we will need them. Incidentally, taking photographs are now easy by using a digital camera or a smart phone and they represent the real figures observed by the researcher. If the journal is asking submitting the research paper through internet then soft copy of the photograph is sufficient though a hard copy is needed in thesis work or where you are submitting the research paper in a hard copy. Where your studies include microscopic examinations, always try to take maximum number of photographs of your microscopic observations, as they may be required while publishing the report or research paper.

CAMERA LUCIDA DRAWINGS

If you are dealing or describing morphology of a parasite, you may present morphological details with the help of a camera lucida drawing (90). This is an old method of describing correctly morphology of a parasite. Here, observations of any parasite structure is made under appropriate magnification of microscope and instead of taking photographs of the structure, you draw it on a sheet of art paper with the help of camera lucida which helps to provide ditto picture, being seen in the microscope, on to the art paper. This camera lucida drawing is important to depict correctly the minute structures of the parasite since photographing of the same will not be good for understanding the detailed description of the parasite.

After preparing camera lucida drawings on the art papers, the important ones are re-drawn on a butter paper with the help of Indian black ink providing magnifications used. Such drawings are acceptable to any research journal.

PART-D

17

Writing the Research Work

What so ever you are doing as a researcher, one important task is to publish the same for scientific fraternity for establishing your view points else no body will recognize you and your competence. Your credentials are judged by the research papers published by you and its impact on scientific community. Therefore, it is must to publish what so ever you are doing to establish yourself in the scientific community and also to advance science. This implies to select a national or international journal for your research publication which has wide scientific circulation so that peers of scientific world may observe your publication. Adversely, the publication may bring you criticism if it does not stand on scientific parameters and you may lose your reputation in your profession.

The first step taken by all the scientists in their scientific career is writing a thesis forward of post graduate degrees. However, thesis has never been considered a solo creation of the student as equally thesis guide has his hidden contribution in shaping a thesis work. There may be many bad or good memories linked with thesis writing as they depend on behavior of your guide, your command on the subject and how you have tackled the critical situations, arose at that times. However, there are certain steps which help you at this learning stage and I will discuss some of them, hereafter.

If you are working in a corporate house or in defense research organization, there are certain limitations of publishing the research work and you are not allowed to publish each and every work; you have to take permission for publishing your work- only academic type work is generally allowed to publish. Nevertheless, in these organizations too, you will have to submit your research report to the authorities with the laxity that the same will not be peer reviewed. Here, your scientific reputation depends on handling certain projects and their outcome, known in the close circuit.

There are certain guidelines which are to be followed while writing a scientific article which is of varied kinds as the same is addressed to different kinds of audiences.

ENGLISH LANGUAGE

English is the language which is used in all scientific publications and correspondence. In recent years, some north Indian journals, particularly published by research institutes or associations, have started providing summary of the paper in Hindi also but whole paper is written in English and so is the case with post graduate theses in science subjects.

Since almost all the universities are accepting the theses in science subjects in English, it's important that you should have good knowledge on English (Even universities in Germany and alike countries are accepting theses in English language, though it is always advantageous to learn another foreign language). Let me clarify that nobody desires you to have master's degree in English, but your standards should be above to high school level, and with this starting knowledge you can carry over efficiently in your scientific career. There is no problem for those coming from English medium but may prove a problem to the scholars coming from Hindi or local medium. It does not mean that an English medium student can write a research article instantly but he is having the advantage of writing English without or less grammatical mistakes.

Though, all the students who have passed BVSc or MBBS and joined post graduate studies have done their undergraduate studies in English medium, generally teachers ignore English mistakes in undergraduate examinations and check only his subject credentials. This is the reason that these students face problems while writing thesis, thesis synopsis or post graduate seminars since there English mistakes are not permitted. So is the problem with our pure science students who are coming from other than English medium.

If you are interested to make research as a career, it must be very clear to command English as a language for writing and communication. However, it does not mean that the person has to develop literary English for writing research reports or articles ; rather it will be sufficient if a student commands grammar, punctuations up to higher secondary standards. The most important part of the language, required, is the grammar which should be perfect with no mistakes. For the beginners, it is advisable to depend on a grammar book of higher secondary standard (91) to understand well about tenses, adjectives, ad verve, active and passive sentence, punctuation etc. It is necessary to increase power of precise writing or abstract writing as this is needed while summarizing or discussing others works. Expansion of ideas is another quality which will greatly help when you will start writing review papers or chapters of the books as it is needed while explaining your idea to the readers.Simultaneously, it is important to increase vocabulary day by day as this will help in your writing. For all the research scholars, it is always advantageous to have one oxford English dictionary,a good Thesaurus (which provides antonyms and synonyms of many words - helps to search exact word for a meaning you have in mind), one science/ medical/veterinary dictionary to increase continuously your English vocabulary (91). Always try and try yourself of writing the work and in this endeavor you may take help of your seniors. In present days, computer may prove your best friend as it has the provisions of checking spelling and correcting the grammar with showing thesaurus of many words. To have perfection, you can add your scientific words into the computer dictionary which avoids checking spelling of that word, every time - neither the computer will show it a wrong word. Indeed, you can improve a lot by putting your writings to the edit part of the computer which suggests alternative words and sentences,other than yours - this also helps you either to follow British or American English which differs grossly in spellings of many words.

With the demand from the students, some universities have allowed the students to submit the thesis in Hindi or other local languages. But experiences have shown that this local language does not work at later stage when the scientist is willing either to communicate with international scientists or desiring to publish his research work in some international journal as most of them are in English language. It must be remembered very clearly that English is the international medium and is only medium to communicate to largest number of international scientists; even if a Japanese or Chinese scientist is meeting in an international conference, the only medium will be English to communicate with them. However, your command in local language is needed when you have to communicate with local people, either during any extension program or organizing a 'kisan-mela' or a public program. Therefore, it is always advantageous if a research scholar can learn as many languages as possible including some foreign languages so that he can communicate with scientific fraternity with ease.

COMMUNICATION SKILLS

Communication skills are very important for making a scientific career though it is not essential for conducting research on a topic. It is required at both levels- writing and oral. The former is required while writing either research article, or any other item on your subject. If you are able to expand the idea and may communicate it more forcibly to your readers, the writing will be much appreciated. You have to expand the idea and make it understandable to the reader. However, it does not mean to include superfluous words as these are prohibited in scientific writings. You may improve in your communication skills by reading research work of senior scientists where you will understand how they expand their ideas and nullify others' observations or find shortcomings in others work.

There is also a need of communication skills during your speech or delivering lectures in your class. This is also needed if you wish to be a good teacher as students always appreciate those who can communicate well to them- how accurately you are able to explain a tough subject in a simple way so that most students may understand them. This is also needed while delivering invited lectures or presenting your research work in scientific conferences, scientific seminars or summer courses. There are some scientists who do excellent research work but are not best in their oral presentation hence fail to attract scientific gathering only on this ground. On the other hand, there are mediocre scientists, who are best in communication, have achieved many laurels only on this quality. Generally, those who have good communication powers, are seen in their later life to join other forum and become an important dignitary in other field- particularly political field. Again, this communication power is needed for participating in international scientific conferences, presenting the data to bureaucrats or other dignitaries who are not well versed with the subject. There it is important how to avoid complicated scientific words and use semi technical words or try to explain the scientific findings in simple non-technical language.

THESIS SYNOPSIS

Where-ever you are planning to take your post graduate or doctorate degree and starting thesis work, every institute has a pre-requisite of asking to submit a research plan of your thesis work. In the traditional universities, there are research committees of different faculties where you have to submit your thesis synopsis, duly signed by your major and minor advisors. This research committee sits on particular day in a year and invite you for an interview regarding your thesis work.

If you have been admitted in a professional college for doing your post-graduate degree, a thesis synopsis is submitted to the research committee etc at the beginning of second year or third semester or after completing your course work. There is a prescribed format for submission of the thesis synopsis. After completing the formalities of informing about department, name of scholar, advisors, you have to provide a short summary of the previous work carried out on the topic, with some important references. This ultimately leads you to say your aims of research work or why you have undertaken the present work or objectives of the present work.

The other important part of your thesis synopsis is a detailed advisory how you plan to carry out your research work. This should provide full information about the materials beings used in your research plan; if you are using any human subjects or animals (domestic or laboratory) whether ethical clearances have been taken from appropriate authorities ; in the matter of biochemical analysis of certain parameters how these will be analyzed with details of the equipments, chemicals and kits being used. Another important point is number of animals used in your experiment or if your total research is depended on experimental trials, how many trials of each experiment will be made and whether these are sufficient for statistical analysis. You have also to provide details of methodology and whether these methods have been used by previous workers or you are planning to develop some new methods to verify your hypothesis and what is the authenticity of these methods. The last part of thesis synopsis is about the possible outcome of your research work and what is its significance or how it will advance scientific knowledge.

On a specified day, you will present your thesis synopsis and have to reply queries. If there are any suggestions, these have to be included in the synopsis and will be followed while carrying your research work. Please remember that without approval of your thesis synopsis by the authorized committee of the college, you are not allowed to submit your thesis hence it is advisable to present the thesis synopsis as soon as possible but by all means prior completion of third semester of your post-graduate studies.

THESIS SEMINAR

To improve research quality of post graduate works,many universities have made it compulsory for the research scholars to present their research work prior submitting the thesis to the university. This presentation is made before a post-graduate research committee or departmental committee or college committee depending on size of the faculty and number of post graduate students. The presentation, now a days, is arranged on power point but a written document is submitted to the committee in this regards. This presentation by the student serves many purposes; the main is that research scholar develops confidence in presenting his work before a scientific gathering and also to find some lacunae in his material methods or results or their interpretation that may be pointed out by the committee or audience or if committee feels that the scholar should carry more work prior reaching to the particular decision. In all these cases, the scholar has to abide by the decision of the committee and has to follow all the instructions while writing the thesis.

While preparing your presentation, you have to mention the previous work carried out by the workers and lacunae in their work which will highlight aims of your research work. Everyone is interested to know what were your materials and methods for carrying your research plan and whether your methods were standardized or were inconsistent. If your work involves some molecular studies, the important items are about mentioning of the equipments, manufactures of the chemicals or biological products and how they are preserved as all these may affect your results. The other aspect is your results which should be depicted in the form of graphs of different forms and also tables, highlighting what statistical analysis was followed by you and whether the results were significantly important or not (please see chapter 16). Interpretation of the results or discussion is the portion which reflects intelligence of the students and how he is able to highlight importance of his work and able to inform the shortcomings of previous work.As this is thesis presentation, the last part of it should contain some important references which have supported your thesis work.

THESIS WRITING

This is the last but an important part in your post-graduate studies. While starting writing the thesis, it should be remembered that every university has published a guideline for thesis writing and every student has to follow these guidelines so that each chapter of the thesis should be as per university norms. If you are unable to obtain the guidelines, you may consult a proceeding year thesis to note the norms or pattern of the thesis. Nevertheless, it must be stated that gist of every thesis is same and the difference between the universities is regarding some minor details like citation of references, some certificates or addition of one or other chapter. Irrespective of any university, every thesis will have following chapters, with the possibility of deletion or addition of chapters like conclusions, suggestions for further research etc.

Front page descriptions as per university notifications.

- Certificates
- Acknowledgement
- Review of Literature

- Materials and Methods
- Results/Observations
- Discussion
- Summary
- Conclusions
- Suggestions for Further Research
- References
- Vita of the Student

The certificates are duly signed by the student and advisory committee except the one which is signed after thesis viva voice. Now a days, the instructions are not to make hard bound copies of the thesis as it is done after completion of viva voice and inclusion of suggestions/ corrections, made by the examiner or the committee.

In general we may say that review of literature and materials and methods are irrespective of thesis results therefore, many students start writing these chapters, much earlier than analysis of results. However, other parts of the thesis depend on your observations or results hence they are written after analyzing the data and getting the results. As mentioned earlier, every university has its own guidelines for thesis writing but basic structure of any thesis remains identical and only difference is in minor details like citation of literature, references part etc which may be corrected as per guidelines. Please remember that thesis writing is giving you an opportunity to express yourself on the topic on which you have worked hardly; it is also important to mention that unlike research papers, there is no page restriction hence you are free to express yourself in right manner.

A word of caution is please do not try to copy passages of other authors /scientists while writing your thesis as than you will never learn how to write a research article (during your post-graduate studies, there is your guide to help you but in later part of your career there will not be anyone to guide you or you may feel ashamed for asking minor details from your colleagues-therefore utilize this initial opportunity fully). Moreover,this habit of copying from other's research paper/thesis may put you in trouble as copying is prohibited under 'copyright acts' and in the past, even some senior scientists have faced this music either due to their careless attitude or mistakes done by their students.

Following is the general pattern in thesis writing which may be followed initially but has to change later as per university guidelines.

REVIEW OF LITERATURE

Review of literature is to inform what and how previous work,related to your thesis topic, has been done by the contemporary scientists.Thus each work of the scientist is discussed,but in summary, what he did, how he did, what were his results and whether there were any shortcomings in his research work. Such review is helpful while discussing your results in the light of previous works.

This review of literature is made on the basis of your searching the scientific literature through referring reference books, review papers, and research papers and has been discussed in details in earlier chapter 14. There are some important factors which one should remember while writing this chapter of the thesis and following guideline will help in this endeavor :

1. The type of scientific literature getting place in your thesis mainly depends on size of the work carried out by the previous workers on the said topic.
2. If there are only a very few references on your topic, you may select references on related topics. For example, if you are working on drug resistance against *Fasciola gigantica* and your search of literature has resulted only a few references, than you may include references related to other fluke infections like schistosomes or amphistomes; again if there are only few references on other fluke infections as well, you may include drug resistance references against tapeworms and if the work is still insufficient, there is no harm to include references related to the nematodes. When the references are less in number, you should include all the references starting from initiation of the work.
3. Just opposite may be true while working on your thesis topic; the topic might be attracting attention of many scientists and there is plethora of information on the topic. In such cases, generally it is the practice only to refer references of last five years or so, beside mentioning some important old references.
    i) As mentioned earlier, you have to write summary of each work which includes all what is required to know for each work. While writing the review, you have to mention the works chronologically, starting from the oldest reference, going down to the newest.
    ii) While reviewing the work, there is not one topic which you are dealing but different topics under different heads and you have to assort your references, chronologically and topic wise. Again,

many universities ask the candidate to review the literature topic wise as well as area wise. For example, if you are working on immunity experiments against schistosome infection in laboratory you have to arrange your review - work done in India, work done in the state, work done abroad. Therefore, you have to assort your references,area wise, chronologically and also may be on schistosome species or laboratory animal species wise, depending on number of references, observed in each category.

iii) Since research work for post-graduate degrees is considered a part of learning of research scholar how to deal a research problem, generally the guide desires that student must learn as many techniques as possible,taking the work in three,four aspects of the research topic; this helps the student to learn and handle many equipments and also help in publication of more papers from his thesis work. And in such thesis, your review of literature should cover each and every topic of the problem. For example, if you are searching hump-sore in Raipur area along with finding intermediate hosts and to undertake treatment trials with a new drug, your review of literature should be divided in following categories - prevalence reports in different states of India, abroad (atleast there are chances of occurring the infection in nearby countries), host species, breeds, sex and age found positive or negative for infection, intermediate hosts in hump-sore or in other Stephanofilarial infection and treatment references related not only to hump sore but all stephanofilarial dermatitis cases, particularly if references are not many. These will also be arranged geography wise for simplification and also if it is required by the university.

4. Try to incorporate all the important and latest references of research problem as same will help you in publishing your work in good scientific journal. For this reason, you should publish your thesis papers as early as possible otherwise, the reviewer of the paper may ask you to up-date the references which at times becomes difficult.

5. Every scientific writing requires many re-writings and corrections and corrections -this is more so when you are learning scientific writing. Therefore, it is advisable to start writing all your ideas which you have build by reading the scientific literature and later refining the writings; in my opinion, if you have re-written any chapter four or five times, then it is worth of showing the chapter to your guide and be prepared to hear his adverse comments on many or in some areas.

MATERIALS AND METHODS

As the title of the chapter suggests, it requires details of three items - the first is materials used in the thesis, other is methods used to solve the research problem and third is your research or experimental plan. Providing details of each item is important as they have the capacity to alter your results. Another important item to be included in materials and methods is statistical analysis to be undertaken.

Interestingly, this is a part of your thesis, which is also un-related to your findings or results thus you may start writing this part,as soon as you have started working on the topic. Indeed, many researchers advocate that students should write all the details of materials and methods while carrying over his research work in his diary ; this helps him afterwards while analyzing his results.

The materials incorporated in research work differ as per your research plan. If you are working on animals for your thesis problem, all the details like breed, sex, age, bodyweight of the animals are needed to mention. If the animals are laboratory animals, the source of their procurement and breeding policy of the animals is required to be known, beside above details. If you are working on susceptibility of an intermediate host, ticks, mosquito, snail, against some infection, details of these creatures are required. Many times, you may procure a snail line or tick stock from a well established laboratory who are working on the topic since many years, you have to mention all these details along with the methods you have used for maintaining them in your laboratories.

If these intermediate hosts have been collected from the wild, it is important to provide details about their procurement and also how you have maintained them in your laboratory before and after the infection. Their taxonomic identification is also important as this is always required while giving authenticity to your work -in such cases, it is always recommended to take help of a taxonomist and same should be mentioned in your thesis.

Now a days, either whole or part of thesis problem is related with some biochemical or immunological work. If you are working on some biochemical parameters of the parasite, you will need particular chemicals and quality or purity of the chemical will affect your results hence it is important to mention details of the chemicals including its manufacturers. If you are working on developing an immunological test against a parasitic disease, you should provide all the details of procuring the antigens, and other reagents. When you are working on plate ELISSA, it is important to mention also details of ELISSA readers. So is the case to provide all the details of other equipments used in your experiments- it may be ultra-centrifuge, ultra-sonicator, stereoscopic microscope, photographic arrangement etc

After mentioning materials, you will provide details of the methods which you have followed in your research work. It is expected that the students should always use standard and well approved methods for investigations. There are many methods which if not standardized will provide varying results hence your results can not be relied upon. Again while providing methods, its important to specify them else it will be confusing. For example, if you are doing fecal examination of humans for detecting schistosome eggs, it is erroneous to mention "feces of school going male children were examined for presence or absence of schistosome eggs"; there are many methods of fecal examination for helminth egg detection and each method has different sensitivity in detecting schistosome eggs and your statement is not revealing which method has been followed by you, therefore, lower prevalence rate in the area may either be depicting a true scenario or it may be because you have not used so sensitive method in your investigations.

Even when you are following a standard method in your thesis, it is recommended to provide a summary of this method. The thesis should provide proper reference from where you have taken the method. Also it is important to mention what changes have been made by you in the prescribed method and what were the reasons for the same.

If you are following any method which is not well recognized in scientific community, you have to describe the method in details. There are certain studies where even one step may change quality of the result. This is presently happening in preparing of fluke antigens. Though the steps followed by the scientist may be identical even a little difference may affect quality of the antigen. In such cases, it is always recommended to provide all the details of your method so that any one can assess quality of the antigen, used by you. In this particular case, even after following each and every step of antigen preparation, if you are not adding protein autolysis preventing chemical, or storing it in ordinary freezer instead of deep freezer or using it after a long time, all will affect quality of the antigen hence your immunological results will be adversely affected. Such variations are also happening in biochemical studies hence great precautions are needed and these should be mentioned while describing materials and methods in your thesis.

There are certain facts in the methodology which are un-certain but may affect your results. Therefore, in such cases, it is important to carry over preliminary work on the subject and mention these facts in your thesis which will improve work quality.

For instance, you are working on prevalence of nasal schistosomiasis in the ruminants and collecting nasal cavities of the ruminants from slaughter house to recover alive *Schistosoma nasale* and as slaughtering of these animals is happening in the evening (as is in Jabalpur) it is important to know if processing of the nasal cavities in the morning instead of evening itself, will affect recovery rate hence will affect prevalence rate in your work. Likewise, if you are collecting fecal samples from wild animals and checking them not for eggs but schistosome miracidia, it is pertinent to know how the positivity will be affected by examining the samples after particular days of the collection (again which is best way to store fecal sample - at room temperature or in freezer or BOD prior their examination). Such questions always arise when ever you are working on alive material and delaying examining the samples. Same question may be asked by the referee when you will submit your research work for publication.

If your thesis problem is regarding surveillance of a parasitic disease in particular geographical area, it is always advisable to carry over a preliminary survey to confirm whether the particular infection is existing in that area or not else your whole thesis program will go in jeopardize and you may have to change it,after confirming that the particular infection is not existing in this area. For example, prior giving me problem on ear-sore for my MVSc thesis problem (92), my professor and guide late Dr SC Dutt confirmed that the infection is occurring in buffaloes at Jabalpur ; though the area does not habitat hump-sore infection - a related stephanofilarial infection,caused by a different species. This preliminary survey is necessary for all the works where it is not confirmed whether the particular biological agent or infection is existing in the location or not.

The last part of materials and methods is to explain your research plan to solve the particular problem. It is important to understand by the examiner if you have missed some steps in your research plan that may affect your results. For instance, if you are making surveillance of a given population for a particular disease, it is important to provide each and every details about the subjects ; in case of animals, species, breed (mention if non-descript are included), age, sex, husbandry practices, geographical area, rough estimate of total population and percentage being examined are to be considered. With regards to human subjects, it's necessary to provide area of investigation, sex and age group, general health of the population (whether you have selected randomly the population or weak or healthy persons were excluded in your survey), education standard of the population (many times, education of the subjects affect hygienic practices); mentioning of season of survey is also important although it is advocated to carry over surveillance work round the year to record seasonal variations in prevalence rate of the infection. Even time of examining the patient may affect your results as shown in filariasis where microfilariae are seen more in the night with difficulty to demonstrate them in blood samples, collected during day time. In short, your plan of work should incorporate each and every details which may affect your results and where examiner or referee may raise question while vetting your thesis or research paper.

While describing plan of work, it will be appreciated if you depict research plan also in a table form as it will provide summary of your research plan. If you are conducting some experiments like blood analysis, never forget to provide details like time of blood collection (which should be consistent), prior or after meals, storage details as glucose concentration will change with storage of blood and serum. The number of repetition of the experiment is essential which is carried out in animal experiments by taking sufficient number of animals and treating them in alike manner along with the controls.

It is essential to support your results by applying statistical methods which will be able to show standard errors, or standard deviations among your data. Whenever you are comparing two or more groups for any variable character, it is important to mention which statistical method will be applied to verify the results. There is no need of describing in detail the statistical methods and it is sufficient to provide reference and summary of the test along with important mathematical formulae.

As we have already mentioned, now a days, you have to clear all your experiments which are based on animals or humans from the ethical committee of the institutes and same should be mentioned in your materials and methods.

RESULTS OR OBSERVATIONS

This chapter is about what results you have obtained while solving your research problem. These results may be of two types - one where you can depict them in arithmetical figures in more better way and the other is where description is better in understanding the results. The results may require both the forms to be included in the thesis. For instance, if you are studying progress of a disease in a given population, it is always better to describe the symptoms which you are observing as the disease is progressing. No doubt, you may also mention while describing symptoms how many persons suffered with these symptoms and how many remained without symptoms and possible reasons for the same. Likewise, if you are observing response of a parasite against certain stimuli, it is always better to describe the responses rather putting them in mathematical form.

Please remember that in this chapter you have to describe or narrate only your results and it is not to be compared with works of others-this is done in discussion. Thus, results is a part of the thesis which either do not have any reference or may contain only a few.

The results are presented with sub headings so that you may explain each and every aspect of your work in discussion. It is necessary to depict your results in different tables and graphs; again variety of graphs may be used for depicting your results. As has been said earlier, thesis does not have page restrictions and this is the learning stage of a student, hence it is desired to depict the results in all forms and also to provide a description for the same.

While presenting your results,you cannot present rough data in this chapter as rough data fails to provide any conclusion to the reader. Therefore, you have to analyze rough data reaching to final figures which are presented in the text as well as in the tables showing percentage figures and marking whether the results were significant or not (and level of significance). It is more easy to interpret the graphs rather than checking data provided in the tables hence graphs should always be presented in the thesis. Though rough data are not given in the results, some guides consider them important for future research work hence ask their students to compile thesis rough data which are shown under "Appendix" (indeed, some important facts of materials and methods are also incorporated under appendix,when it is considered important).

It will happen that some of your results may be alike of previous workers and may be insignificant to the scientific community. But you may obtain some results which are important, different from general perception and may require further investigations. And these results should be highlighted while writing your thesis. For example, efficacy of praziquantel was analyzed in porcine schistosomiasis by two methods - by recording number of schistosome eggs per gram in pig feces and by counting the blood flukes after necropsy of the animals (2). The fecal examination method revealed that the drug decreased fluke eggs in pig feces to a significant level hence it was expected that necropsy of the animals will also reveal significant reduction in blood fluke number. But contrary to this expectation, no significant blood fluke reduction was observed. Thus, this result becomes important for future work as it is suggesting that you cannot correctly judge efficacy of a drug against schistosome infection by egg count in host's excreta- the only way of judging parasitological efficacy of a drug during ante-mortem stage. This observation requires further exploration - whether this is true only in this host-parasite model or only for certain time of infection or only for this drug ; in last eventuality, the drug might be influencing significantly egg production by the flukes but without killing them in a significant manner.

Please remember that many examiners scrutinize results section of thesis minutely as according to them this is a reflection how much work has been done by the student for his thesis problem and whether he is able to present his results correctly.

DISCUSSION

This is an important part of your thesis and really reflects intelligence of the students ; many examiners who are well versed with thesis subject read carefully discussion of thesis to reach any conclusion about quality of the thesis. In fact, discussion reflects many facts- it is about how you are able to protect your results in light of other works ; how you were able to find shortcomings in material and methods of previous workers hence erroneous results or how you have improved your materials and methods to obtain more accurate results or how interpretation of previous workers was wrong about their results or how your results raise questions about general perception on a topic in scientific world. Obviously, for discussing all these matters, a deep knowledge of the subject is required and that's why I have said that discussion reflects intelligence of the student.

In some old theses, you will note the type of discussion which leads us no where as it contains "our results are in agreement with those of .... And our results are in difference with......" Actually, this is no way of discussion unless you provide reasons for differences of these results.

As I have referred above, discussion needs your thorough study of previous works meticulously not one part of research paper but all three important parts i.e. materials and methods, results and discussion and you have to compare each of these from yours to find correct assessment of previous as well as your results.

Nevertheless, it is the results which make thesis important. If you have worked seriously on your research problem,there are all chances to record some important findings which need to be highlighted and which generates back ground for future research work.

Here, I am citing some examples from my post graduate theses where some new findings were observed which are important to carry over further research work; remember that pursuing such fields mostly results in some unexpected findings which become milestones in our advancement of science.

The previous works on mouse- *S.incognitum* model showed that though albino mouse develops the flukes to adulthood with presence of its eggs in liver and intestine, yet it does not pass eggs in the feces hence this parameter could not be taken in any experimental work employing this model. However, meticulous work of my post graduate student Ninder Kaur (93) revealed that mouse do excrete these eggs though not before 40 days of post infection and when more specific coprological technique (acid-ether) is employed. As this was an important finding of my post-graduate student, it was mentioned with due justifications in results and it was discussed in details in discussion, after taking into consideration all the facts.

In the Ph.D work, Samidha Gupta (94) infected Barbari goats simultaneously each with 2000 cercariae of *Schistosoma incognitum* and 2000 cercariae of *S.spindale* hence it was expected that there should not be significant variation in number of adult flukes, developed, of the two species. However, necropsy examination of these goats,at different time intervals, recovered just half the number of *S.incognitum* in comparison to those of *S.spindale* ; death of a goat on 605 days post infection witnessed presence of only *S.spindale* with no specimens of other species. This work was simultaneously followed with fecal examination of the goats continuously, at weekly intervals, for counting the eggs of the two species. Again, it is a fact that *S.incognitum* female possess only one egg in its uterus while there are many in *S.spindale* female giving the probability of higher number of eggs of the latter even where female number of the two species are same. Surprisingly, it was not so and by counting the eggs in the feces, no one can predict that number of *S.incognitum* will be just half or less than that of *S.spindale*. And these are interesting findings which need further exploration. Therefore, this work was highlighted in the thesis while mentioning results and discussion.

Another example I wish to mention is about undertaking chemotherapeutic trials with Praziquantel in experimental porcine schistosomiasis using *S.incognitum* infection (95).The efficacy of the drug was judged by using four criteria- hemoglobin concentration, body weight of the animal, fecal egg count and number of flukes recovered from treated and untreated animals. The results showed improvement by chemotherapy in hemoglobin concentration and body weight with reduction of fluke eggs, all at significant levels. Thus it was presumed that there will also be a significant reduction in *S.incognitum* number but this was not so. It was obviously an important question how there was clinical recovery when fluke numbers were not reduced significantly ? Whether reduction of fluke eggs was sufficient to influence health of the animal to a significant level ? How much will be the improvement when the drug is effective even in killing the flukes to a significant level ?As these results are practically important, these were found special mention in results and discussion in the said thesis.

Like any other chapter, it is advisable to write first all the details of your results and how you weigh them in light of previous works or for future work. Afterwards, it requires refinement and addition of specific references where ever needed. Like review of literature, discussion is another chapter which requires addition of many references and the pattern of their inclusion should be as per guidelines of the particular university.

SUMMARY

This part of thesis is written without any reference ; it is not only summarizing your results but requires more details. Indeed, summary should provide in short about previous work, aim or objectives of your thesis,how you tackled this problem or materials and methods, important observations or results which are advancing our knowledge and why previous workers failed to note such results or reasons for the differences and importance of your work in scientific arena. The summary is an important document as this is referred at many places where thesis is cited but not available. Summary does not require any references and you should continue to write and re-write the summary keeping in mind that it is an important document.

CONCLUSION

Conclusion is shorter than summary and should not include details,as prescribed in summary. Thus, it omits many details which are mentioned in summary. Indeed, some universities do not require conclusion and consider it rather redundant.

SUGGESTIONS FOR FURTHER WORK

This is a part of thesis which has been included by some universities, under recent changes. This is about new results, obtained by you and if they are worth for further investigations.

REFERENCES OR BIBLIOGRAPHY

This is an important chapter that requires inclusion of all references that have been cited in your thesis. This chapter reflects how meticulously you have dealt your thesis and many examiners scrutinize this part of thesis mainly for finding common errors,made by the students. These are - omission of the references in the list that have been cited in thesis anywhere, sequence or chronology of references and whether they are as per standards followed by your university and mistakes made in citing particular references; this may be in omitting initials of some workers or name of the worker (particularly in multiple authorships), use of comma, year of publication, title of the paper, name of journal (a common mistake is made in abbreviations of journals, if that is needed) mistakes in italic,bold or normal formats for volumes of the journal, and issue number if required and finally page numbers; generally both first and last page of the citation is required in theses. Examiners are also particular if students have left any important reference on the subject. This reflects the student has not searched the literature properly ; alternatively the left out reference may be important in assessing your thesis work as results of the two are wide apart - in such happening it is difficult to protect your results and really it reflects intelligence of the student at such time.

There are various ways of citing the literature in thesis and for that matter in any scientific writing ; every research journal or university follows their own pattern of citation. There are differences in citing references of research papers, technical reports, seminars and books. To avoid repetition, I have mentioned these types while describing writing of research papers and the same will be followed as per your requirement in thesis writing.

APPENDIX

If you consider that your rough data or some other information like formulations of solutions, some equations etc are important for getting the proposed results and also for future work, you may add this information in appendix so that these may be consulted later while carrying or scrutinizing your work. For example, if you have examined viscera of a number of animals for presence or absence of particular parasitic stage, you may provide animal wise data in the appendix which is not possible in results.

VITA

This is the last part of the thesis which has been added afresh by many universities. This is one page information regarding personal and academic information of the student. Here you should mention your service experience, distinctions in examinations, your hobbies, games and research experience and publications if any. However, the information should be provided in neutral way without reflecting any biasness about your achievements.

COMPACT DISC

Many universities have started asking students to submit the thesis in a CD form also as it is easy to assess it on the computer by subsequent students or researchers. Even if it is not mandatory by the university, it is advisable to have a CD for your personal usage as it greatly helps you in writing the research papers, from thesis work. It is possible that loading of theses on university's website may become a normal procedure in future as has been done in many foreign universities. This system strictly avoids repetition of research work while taking fresh thesis problem, make retrieval of any part of the thesis quickly and may be scrutinized easily for plagiarism between the two theses.

WRITING A RESEARCH PAPER

Studying research papers of eminent scientists is the best way of learning the writing of a research paper as this act imbibes many basics which are difficult to narrate. It is always advantageous to develop habit of reading full research papers from important journals and do not depend only on the abstracts of the research papers. However, a word of caution for the beginners is never to copy any part of the published paper, review paper, reference book etc, (may be a graph, table, photo, or even text) in your to be published paper as it may lead to great trouble as it is illegal and you may be prosecuted for copying any part of the published work without permission. It should be clear that copying some part of published papers may be ignored in your thesis writings as this is not considered by many as published work and generally the seniors take a lenient view in such matters but this may not happen with research papers. Or it may happen that you are proved culprit of plagiarism but pardoned by the author or publisher, even then this is disgraceful and may prove a black spot in your future career. Please remember this type of copying is different than discussing your findings in the light of previous works which you have to do for support or otherwise of your contention on a certain topic.

In attempting to write a research paper, knowledge of English is important but scientific writing does not require literary language and be devoid of similes, examples etc ; It should be simple, without excessive words and restricted only to the words which are necessary to communicate the message to the scientific community. Earlier, the research papers were written in passive voice but now active sentences are preferred but in past tense (except universal truths).

Each journal prescribes guidelines to the contributors where instructions are given for titles under which you have to frame your research paper and also format for writing the references. Following the instructions is essential prior submitting the paper to the particular journal else Editor may reject the paper,simply on the grounds that "it is not as per format of the journal". Please remember that prior submitting the article to the journal, you have to write the paper several times and computer has made this exercise easiest in its present form.

The best policy in writing a research paper is first to jot down your thoughts on the paper or on a computer file one by one without caring for other details. Your writings should then be categorized in different parts of the research paper. These ideas and findings should be restricted only to the area of your work and do not try to comment on the topic which is irrelevant to your work as referees do not like comments where you have not worked. All the research papers, even short notes, have to communicate in the form of an introduction, materials and methods, results or observations, discussion and summary or abstract and finally references. It is a different matter that short notes do not depict all these headings in the paper (some journals even are mentioning these headings) but the author has to mention all these in their short notes too. Therefore, without thinking whether the paper is to be submitted as full research paper or as a short note or identifying the journal, one should start writing all relevant matters and then may incorporate other details like references or data of the work, either related to your work or of others. You have to provide following information in every research publication :

TITLE OF RESEARCH PAPER

This is the starting point of your research paper. Initially, you may select 2-3 titles of the research paper and finality may be decided afterwards. The title should not be too long and should communicate what you are willing to present in the research paper ; the reader must be clear whether you are dealing taxonomy, or life cycle of a parasite or epidemiology or diagnosis or chemotherapy of a parasitic disease or you are presenting a unique method of controlling a parasitic infection or this is a unique finding of a parasitic species in some new geographical area.

The name of workers or researchers is next to be incorporated with their official linking i.e. where research work was carried out. Here neither degrees are mentioned nor positions of the researchers are listed. The sequence of research workers in the research paper is important for giving credit of the work ; this is in descending order from first author- the main scientist who has contributed mainly in research paper. Importantly, the first author is responsible to answer if any dispute in reporting arises. If the research paper is the outcome of thesis work, the first name and due credit goes to the research scholar who is entitled to keep his name first in such publications. The second name generally is of the major guide though at times,he may desire to replace it with minor guide who has contributed most to that part of the work. It will be a bad commentary to incorporate more than four names in thesis work. Where the research work is a product of some research project, a research fellow cannot claim his name being put first or second in the research paper as this is the result of technical planning of the principal investigator who is entitled to get the credit for the work being carried out but name of research fellow should also appear in the paper.

The first page also covers name of corresponding author to whom editor will correspond in future regarding any queries related to the paper and correcting proofs of the paper. It is now customary to provide e-mail and cell phone of corresponding author also for quick contacts. Generally, this is the major guide of the thesis or principal investigator who represents as corresponding author. If a research scholar or any other researcher has left the institute of the work, the present address is provided for this scientist in a foot note.

KEYWORDS

Almost all the journals are asking to provide 5-6 key words in the research paper. These are used by abstracting journals and other software media for retrieving the paper and also helps you to find related research references. By searching any of these key words, your research paper may be traced. Key words may be related to geography (India, Bihar), name of parasitic disease (sarcocystosis, trypanosomiasis, schistosomiasis), parasitic species (*Schistosoma nasale, Trypanosoma evansi*), host (tiger, dog snails), or diagnosis, epidemiology, immunodiagnosis, feces, control, new technique etc.

INTRODUCTION

This is the beginning of the research paper which introduces the readers with the research paper. Thus introduction is to communicate about the present problem and what is the opinion and findings of previous workers over this problem and how the present workers plan to deal the problem. Earlier, it was customary to review most of the literature, related to the problem, in the introduction but now the trend has changed and the authors are supposed to critically review only latest four -five references drawing their lacunae or how differently the problem may be tackled. Perhaps, this change has occurred first for curtailing space, research papers are now dealing still shorter subjects than previous ones and also as important national journals are not interested to publish repeat works without some new ideas.

ABSTRACT OR SUMMARY

Where journals are publishing summary of the research paper, it is provided in the last of the text but where abstract is inserted in place of summary, it is provided prior introduction and key words. Many abstracting journals publish these abstracts in their journals without any change, therefore, abstracts should be written keeping this fact in mind that these are the words which most of the scientists will come across and will be a source to know about your work. There is a word limit (100-250) in summary/abstract which should be without citation of any reference. Summary is neither description of material and methods nor results or discussion but a combination of all without putting details of any of these and conveying the most important facts of your research work so that any reader may grasp what and how you did the work, what were your findings and what is its importance in present context.

MATERIALS AND METHODS

There are two different matters which are dealt here. First is about your materials where you have to mention type and quantity of the material that is used in the experiment. This may be collection of blood samples from slaughter houses, or experimental animals or domestic animals or fecal samples with details like time of collection, preservatives if used etc or details of animals like sex, age, breed used in the experiment or alike matters. While you are mentioning materials, please get assured that it contains detailed information which may influence your results else the future workers may raise doubts about the results on this ground alone. For instance, if you are describing a coprological method for diagnosis of a parasitic infection and has not mentioned time of collection of fecal samples, the future workers may raise questions on your results as later it is suspected that egg or cyst excretion is influenced by the time i.e. may have more concentration in night or morning feces. If you are making biochemical, hematological or enzymic studies, it's important to mention time of material collection, preservatives used, storage process and time of animal feeding, beside other details. These details are also important if your experiment is dealing with clinical observations of a parasitic infection in a host species as not only the latter but former parameters may also influence results of the experiments. In your biochemical experiments, it is important to mention quality of the chemicals, enzymes, antibodies along with name of the product companies as many times quality of these chemicals are responsible for different results. In the molecular studies, it is also important to mention strains of the parasites, hosts and source of their procurement as all these may affect greatly your results.It is also important to mention type and mark of the equipments such as ELISA reader, deep freezer, centrifuge machine, ultra-sonicator etc.

The second part in the materials and methods is describing the methods or methodology of the experiments. Here references will suffice when you are describing a well known methodology but do not forget if you have made changes in the prescribed methodology which might have influenced your results. For instance, even if you are employing a well known technique of salt floatation method of fecal examination, whether you have filtered the feces and type of sieves used for the purpose,all will affect results hence its important to mention details of the sieves in such case. If the methodology is developed by your laboratory, its important to describe the same in precise way with factors which may cause variation in the results so that any other laboratory can follow the same. If this technology has been described in your previous publications also, its important to mention those references also. Mostly, it happens that we modify a previously described technique as per local suitability or availability of the materials ; in such cases, it is important to provide details of this modification with citations of original references of the technique. For example, we followed a perfusion technique of Smithers and Terry (96) for recovery of schistosomes from albino mouse. However, these authors used a peristaltic pump for infusing citrated saline in the animal which was not available in our laboratory hence we tried successfully an automatic Pippeting machine. Secondly, they used a separating funnel for collecting perfusate and searching blood flukes from bottom fluid under a stereoscopic microscope. This whole process appeared to us cumbersome, time consuming and we replaced it simply by filtering the perfusate with a black muslin cloth and inverting the same in a Petri-dish (already containing some fresh saline) which resulted in transfer of all the blood flukes in the Petri-dish ; nevertheless, prior describing this technique we have checked several times,while filtering, if muslin cloth was able to retain all the schistosomes including the smallest ones and whether agitation of the cloth in the Petri-dish transfers all the blood flukes into the Petri-dish. Only after confirming these points,we described the technique in our paper (97) providing original reference of Smithers and Terry and modifications made which simplified the technique to be used in any less developed laboratory.

If your research work is about prevalence of sarcocystosis or hydatid infection in canines, it is important to categorize your dogs into pet animals and stray dogs; it must clearly be investigated whether the locality was having slaughter houses or meat shops (with species of flesh) and assess of the dogs to these places as all these factors will greatly influence prevalence of the infection in the canines. In summary, it must be remembered to mention the factors which are variable and may influence the results and that you have taken care of these factors while employing materials and methods in your experiments.

Use of International Units

For better understanding of the research work globally, it is decided in 11th general conference of weight and measurements of the international organization for standard (1960) that all the scientific works should be reported using international measurement units so that it may easily be understood- to use coherent units and decimals and sub-decimals of these units. Metric system has been followed by the scientific community and prefix are used in descending orders as follow : mega (M 10^6), Kilo (K 10^3), hecto (h 10^2), deca (da10^1), deci (d 10^{-1}), centi (c 10^{-2}) milli (m 10^{-3}), micro (u 10^{-6}), nano (n 10^{-9}). Thus the measurements for weight are depicted in kg, gm, mg, ug, for length in meters, cm, mm and micron, area in square cm or meter, volume in cubic cm or meter, rain falls in cm per year or cm during rainy season, climate temperature in Centigrade (though body temperature is still given in Fahrenheit); whenever describing any geographical area, always mention its location by latitude and longitude; when ever you are working on blood or biochemical parameters or enzymes, they are presented in international units. Again there are fixed abbreviations for each parameter which have to be used in the research paper.

RESULTS OR OBSERVATIONS

This is most important part of your research paper as it communicates your findings on particular topic; these are your results or observations which are referred for many years by the scientists if they withstand test of scrutiny with passage of time. Generally we are using the term observations in place of results where it is difficult to quantify each and every aspect of our results. For instance, its better to describe morphology of a parasite under observations as it gives you leverage to mention other details also or you may deal clinical findings under observations or ecological conditions for survival of the snails. You may describe epidemiology more appropriately under observations.In short we use observations where description is more important than mere data. Term results is used where our observations are presented in a better way by quantifying manners or where future scientists will prefer to refer our results citing the data. But here also data are confined into one or two graphs or tables which will high light your results and in the text you are permitted only to use significant data. If results are associated with the photographs, these should be described summarily in the text ; obviously, it is must to refer all the tables, graphs and photographs in the text.

Please check if you may incorporate all your data in the text thereby avoiding tables as journals are not preferring tables where they may be deleted or may charge for its inclusion. Though results are most important part of the research paper, you are not permitted to incorporate raw data or which carries no meaning. If you are presenting one year observations on fecal egg output, or body weight loss or gain during the period or alike data, you are not supposed to provide day to day data and you have to concise the data either on fortnight or monthly basis although mentioning if any significant deviations were observed for any period; instead of presenting such a long data in tabular form, many journals prefer them in graphic form which will save space and will also convey instantly its message.

It is important to state results according to the categories of the animals to which they belong. If you are dealing prevalence of sarcocystosis in dogs, your results should describe its prevalence in dogs having free access to slaughter houses or meat shops *vis a vis* which do not have such opportunity along with sex, age and breed of the animal ; then only it will convey a fruitful message but if you have not analyzed your data in this context, your results will be confusing and will not convey the correct message. The statistical analysis between these two groups and between different age, sex and breed of the animal will communicate importance or otherwise of each factor in determining prevalence of the infection in this host species. Likewise, whether animals are allowed for grazing or not and its details are important in studying prevalence of fluke infections. If these data are not analyzed properly, it will not convey the correct message. In working on immuno-diagnosis it is important to mention source of antigens and its preparation and also whether the test was performed in experimental or field infections ; in these cases also, its better to compare the results with most sensitive parasitological test to prove efficacy of immunodiagnostic test (as unlike to viral or bacterial infections, immune-diagnosis is still not standardized and varied results are reported by different laboratories). If you are reporting drug or vaccine experiment, its better to analyze its effect on every stage of the parasite i.e. immature, mature, male and female worms as the drug may be more effective on immature or female worms rather than whole group of the parasite ; absence of such analysis may minimize importance of your work when future studies are carried out taking into consideration of these facts. Likewise, vaccine experiments will also require all the details to show whether vaccine has minimized incidence or severity of the infection in a given population.

DISCUSSION

This is also an important part of your research paper. While most of the scholars consider writing of materials and methods and results being simple, they consider writing discussion a tough task. Perhaps, because materials and methods incorporate what one did and results are the facts which are before them while discussion is a creativity to be generated not only from materials, methods and results but also by analyzing the work already done by previous scientists.

In fact, discussion means where your results stand *vis a vis* those of previous scientists and what your results convey for future thinking. However, the most absurd way of writing the discussion is 'our results are in agreement with so and so' and 'our results are contrary to the findings of so and so.' Such writings simply show lack of interpretation power of the scholar or they do not possess deep knowledge of the subject. Indeed, the reader of the paper aspects from you the reasons of similar or different findings from previous scientists. If you have analyzed earlier works correctly, the answer lie in their materials, methods or in observations. For instance a significant difference in prevalence of the infection of yours and previous workers might be because of difference in geographical area surveyed, age, sex and breed of animal or animal husbandry practices followed in the two cases. Fluke infections will be significantly low in those ruminants which are not allowed grazing ;

dogs would have lower percentage of sarcocystosis due to low or no assess to meat shops etc. Effect of the same drug in identical parasite species may be different because of use of different host species, or duration and dose of treatment or as you have analyzed its effect both in mature and immature worms or male and female parasites. We were able to demonstrate excretion of eggs of *Schistosoma incognitum* in the feces of experimentally infected albino mouse while all our previous workers reported negative results. And reasons for the same were duration of fecal examination and more sensitive diagnostic method employed (57) ; and this was not one time observation but remained true as confirmed by subsequent works when the feces were examined by acid-ether method after 40 days of experimental *S.incognitum* infection.

Even with your best efforts, if you fail to understand the reason for wide difference of the results of yours and of previous workers, there is no harm to mention the same with the need of further research to settle the records right.

It must be clear that discussion must include all the new references of similar work of previous workers and if they are many you may restrict their number 8-10 but revealing that you have reviewed the literature thoroughly; at times the research papers are returned with the remarks that the workers have not reviewed the literature properly. It is always advantageous to review the literature prior starting research work, yet some scholars begin searching them while writing the research paper. The greatest disadvantage of such misadventure is that you may come to know that your work has already been done by previous workers with much more parameters than yours thereby diminishing the chances of publishing your research work in an important national or international research journal.

Many post graduate scholars do not care publishing their thesis work as soon as it is finished and plan to publish it after two -three years, when they are settled somewhere ; at this time it becomes difficult to publish due to absence of latest references with the risk that during this ensuing period, others might have published similar research work. Therefore, it is always advantageous to publish your research work, thesis or research project, as soon as its work has been completed.

REFERENCES

This is the last part of the research paper which incorporates all the research references which have been mentioned, any where, in the research paper - introduction, materials and methods, results or discussion. Citing the references in the research paper confirms that what so ever you have dealt in the research paper carries scientific meaning, it's not a fiction but your work, depending on some scientific base and in support of this you cite the work of others and at last provide details from where this citation is taken so that reader may retrieve any reference,if he desires so.

References are cited at two places in any research writings - book chapter, review, research paper or technical report. The first is in the text and other is citation at the end of the research paper. There are different ways of such citations and each journal/publisher prescribe the exact procedure for its citation and it is best to consult these guidelines prior finalizing the paper.

Citation in text

a) In traditional way,the reference in the text is cited with the name of the scientists and year when this work was published. This published work may be in the form of a research paper, review paper, book chapter or a technical report. However, names of the scientists are restricted either to the sole scientist of the research paper or two scientists if there are only two authors of the paper. It is only surname of the scientist that is mentioned without initials of the name. In case, the cited paper has more than two authors, it is cited using name of first author and suffixing it with *et al* indicating there are more than two authors of the research paper. The year of publication is cited generally in brackets though it may be without bracket, if writing desires so. If two or more papers of the same authors in one year are cited in the research article, they are differentiated by suffixing the year with alphabets- 1994a,1994b

   Example :

   i) Kohli (1991) and Kohli and Agrawal (1994a, 1994b, 1996) made extensive observations on cercariae of *S.spindale.*

   ii) Gadgil and Shah in 1952 demonstrated existence of urinary schistosomiasis in Gimvi village, Ratnagiri district of Maharashtra State, India, incriminating a different snail than *Bulinus* sp.

   iii) There are six sets of sensory papillae and 22 epidermal cells are arranged in four tiers (Chauhan *et al.*, 1973).

b) There are some journals (e.g. *Proceedings of National Academy of Sciences,* India Allahabad) which have changed citation in the text and requires numbering of the references. Here the references are numbered, first come first, and cited in the text by serial numbers whereas in the References, they are mentioned as per their serial numbers ; here neither alphabetical order of first author nor chronology of publication is followed while mentioning them under reference part of the writing.

Example : Cercarial dermatitis is found endemic in rural India from South India to North, East to Central India where ever studies have been undertaken [1-4]

In the Reference part of the research paper, it is mentioned as follows:

i) Narain K, Mahanta J, Dutta R and Dutta P (1994).

ii) Narain K, Rajguru S K and Mahanta J (1998).

iii) Agrawal MC, Gupta S and George J(2000).

iv) Agrawal MC (2003).

c) There are other journals who are numbering the references in the text according alphabetically as per authors surname so that numbers are not as per their first appearance in the text but as per alphabets of the authors ;thus, in the text first number may not be one (if it does not start with AA) but may be 9,depending on where it stands under References. (in the above case, reference number 4 and 3 will be 1 and 2 and others will also be changed).

Example : Sewell (9) published a monograph on Indian cercariae while Bhalerao (2) published another monograph based on updated information on helminthes of Indian livestock.

The advantage of such citation is easiness of searching the citations and helping in cross-reference work.

Citation in Reference part

Where journals are accepting name of the authors,instead of numbers in the text, the references are given in the last alphabetically and chronologically with name of all the authors, first surname than initials of their names, separated other name with comma (generally journals have deleted comma in between surname and initials of the same author), year of publication, title of the research paper, name of the journal with volume and page numbers. For same author, the research paper in his solo name comes first, followed with papers with two names and then three names and so on. There are changes as per journals and one must mention the references as per guide lines provided by these journals.

Though most of the journals are asking name of the journals without abbreviations, there are still some journals (e.g Indian Veterinary Journal; Journal of Veterinary Parasitology) which are accepting standard abbreviated names of the journals which you have to search since correct abbreviations are must for publishing the research papers (see chapter 14). We are providing herewith examples how some journals are following their norms-each parameter is important for citation e.g. whether year of publication is under bracket or not,volume number is italic or bold:

Research Paper citations

Indian Veterinary Journal : The journal do not require Title of paper and only first page of the paper is mentioned. Name of journal is written in italics while volume is given in bold pattern

Ahmed, Z (1961) ..... *Indian Vet J.* **38** : 257

– – – – – (1962) ...... *ibid* **39** : 120

If papers of same authors are given, no need of repeating the name in second paper, instead stretch a line indicating same authors. When journal name of subsequent research paper is same mention '*ibid*' instead of writing name of the journal.

Journal of Veterinary Parasitology/ Journal of Parasitic Diseases

Chandler, A.C.,1926. A new schistosome infection of man with note on other human fluke infection in India. *Indian J.Med.Res.,* **14** : 179-183.

Though name of the journals are abbreviated, volume is not given in bold format ; further both first and last page number of the publication is provided.

Journal of Parasitology and Applied Biology's pattern is same as above except asking volume in bold letters as follows :

Chandler, A.C., 1926. A new schistosome infection of man with note on other human fluke infection in India. *Indian J.Med.Res.,* **14** : 179-183.

Indian Journal of Animal Sciences : This is an important journal of ICAR which asks titles of research papers; full name of the journals is given in italics while its volume in bold :

Sadun E H, Von Lichtenberg F and Bruce J I. 1966. Susceptibility and comparative pathology of ten species of primates exposed to infection with *Schistosoma mansoni. American Journal of Tropical Medicine* **15** : 705-18

Many foreign journals like Annals of Tropical Medicine and Parasitology, Transactions of Tropical Medicine and Hygiene are asking full title of the journals,instead of their standard abbreviations. Again, some journals are providing year of publication at the last of citation (e.g. Patnaik, M.M. and Pande, B.P. Notes on the helminthic infestations encountered in one month old buffalo calves. *Indian Veterinary Journal* **40**: 128-133 (1963).

Citations of non referred articles

2. There may be certain citations where you have not consulted the original paper but have cited from some other reference; here the citation in the text is made of original worker e.g. Cobbold (1882). But following procedure is followed for mentioning the same in Reference.

Cobbold T.S. (1882) Cited by Baugh S.C. (1978) A century of schistosomiasis in India : Human and animal. *Rev. Iber Parasitol* **38** : 435-472.

Citation of unpublished articles

Some papers which have yet not been published but they have been accepted for publication in some journal, you may refer such papers also but without giving volume or page number. Here only accepted papers and not 'submitted' papers are cited. Obviously, the latter are excluded due to uncertainty of their publication.

Kalha R, Agrawal N and Katoch M (2010) Epidemiology of bovine Paramphistomosis in Jammu. *Journal of Parasitic Diseases* (in press).

Citation of personal communication

Some times, a scientist or your colleague might have communicated to you through correspondence some important fact on any topic or his experience over the subject ; there is no research publication which can support it but you wish to refer this fact in your research paper, which you may mention citing the reference only in text of that scientist by writing -personal communication.

For instance, we have observed that goats were negative for *S.nasale* in Balaghat district and same has been observed in 2012 by Moghe in Bhandara district but without publishing the work. Then to substantiate our claim we will mention in the discussion "Goats were negative for snoring disease in Balaghat district and same has been observed in Bhandara district as well by Moghe in 2012 (personal communication or Moghe 2012-Personal communication). Such references are mentioned only in the text.

Citation of online journals

There are now research journals which are online and may be assessed by any one without subscribing them. At times, these journals may contain some important research paper which you wish to refer in your coming research paper; in such circumstances, the reference is given in such a way so that any other scientist may open the site of the journal and paper and may read the desired research paper:

Huyse T, Webster BL, Geldof S, Stothard JR, Diaw OT, Polman K, Rollinson D. (2009). Bidirectional introgressive hybridization between a cattle and human schistosome species. *PLoS Pathology* **5**(9):e.1000571

Agrawal M C and Rao V G (2011). Indian schistosomes : A need for further investigations. *Journal of Parasitology Research,* Volume 2011, Article ID 250868

The above is the method of citing online research paper but there are various ways of retrieving the particular research paper. For example, in latter case :

a) You may access the journal by following its URL: http://www.hindawi.com/journals/jpr/

b) By following its bibliographic information : DOI:10.1155/9582 This DOI means digital object identifier which is a unique number assigned by an international agency to provide a persistent link to its location on the internet.

   Thus by logging by either way, you will reach to the journal of Parasitology Research where you may assess any research paper.

c) However, if you wish to access the particular research paper you have to provide name of journal and then its ID number i.e. Journal of Parasitology Research Article ID 250868. Thus without knowing details of the authors or title of research paper, you will reach directly to the specific research paper.

Thesis citation

The pattern will change according to the journals and may be :

Agrawal, M.C. (1973). M.V.Sc. Thesis, JNKVV, Jabalpur.

Agrawal M.C. (1978). 'Studies on heterologous immunity in schistosomiasis'. Ph.D. Thesis, Jawaharlal Nehru Krishi Vishwa Vidyalaya, Jabalpur.

Mishra A (1991). *Development of Schistosoma spindale and S.indicum in the laboratory and their diagnosis.* Ph.D thesis. Rani Durgawati University, Jabalpur.

Banerjee, P.S. (1988). Comparative efficacy of different diagnostic methods in determining the prevalence of schistosomiasis in cattle and buffaloes. M.V.Sc. Thesis, Jawaharlal Nehru Krishi Vishwa Vidyalaya, Jabalpur.

Book citation

Books may be two types, from where citations have been taken in the research article. One type of book is where it is written by a single or multiple authors while the other is where each chapter has been contributed by different authors and whole book is edited by some other authors. While mentioning book reference in the text of the paper, the name of the author of the chapter (edited book) or of the book is given with the year of book publication but citation in the Reference section is different and is as follows :

Agrawal M C (2012). Schistosomes and schistosomiasis in South Asia. Springer (India) Pvt Ltd, New Delhi Page 351.

Bhalerao GD (1935). Helminth parasites of the domesticated animals in India. Scientific monograph No.6. Imperial Council of Agricultural Research, New Delhi,p 365.

Dutt SC. (1980). Paramphistomes and Paramphistomiasis of domestic animals of India. Punjab Agricultural University, Ludhiana.

Subba Rao NV (1989). Hand Book : fresh water molluscs of India. Zoological Survey of India, Kolkatta.

Agrawal M C (2012). Schistosomes and schistosomiasis in South Asia. Chapter 2 The Schistosomes. Springer (India) Pvt Ltd, New Delhi Page 7-50.

Agrawal M C (2003). *Epidemiology of fluke infections.* In (editor Sood M L) Helminthology in India. International Book Distributor, Dehradun. Page 511-542.

Agrawal M C (2003). Epidemiology of fluke infections. In *Helminthology in India*. Ed, Sood M.L. International Book Distributor, Dehradun. Page 511-542.

Chowdhury N, Sood ML, O'Grady RT (1994). Evolution, parasitism and host specificity in helminths. In Chowdhury N, Tada I (eds) Helminthology. Narosa Publishing House, New Delhi, pp 1-33.

Taylor, M.G. (1987). Schistosomes of domestic animals ; *Schistosoma bovis* and other animal forms. In *Immune responses in parasitic infections: Immunology, immunopathology and immunoprophylaxis. Volume II : Trematodes and Cestodes* (edited by Soulsby, E.J.L.) Boca Raton, Florida, USA, CRC Press Inc. 49-90.

Technical report citation

In the text, the procedure for citing reference of technical report is identical as for other research publications. In the reference part, the citation is given with address where the report is submitted (sponsoring agency) and not where work is done ;

Alwar V.S. (1974). Final report on All India coordinated research project on "Investigation into the factors governing the epidemiology of nasal schistosomiasis in bovines and its control in different field conditions. Indian Council of Agricultural Research, New Delhi (e.g. the work was done at Chennai while report was submitted to ICAR, New Delhi).

Agrawal M.C. (2000). Final report "Studies on strain identification, epidemiology, diagnosis,chemotherapy and zoonotic potentials of Indian schistosomes" National Fellow Project, Indian Council of Agricultural Research, New Delhi.

Mishra A K (2004). Final Report NATP MM sub project "*Diagnosis of parasitic diseases of domestic animals*". Indian Council of Agricultural Research, New Delhi.

Agrawal M C (2002). Second Annual Report of Cooperating Centre, NATP MM sub project "*Diagnosis of parasitic diseases of domestic animals* ". College of Veterinary Science and AH, Jabalpur.

Conference, training citations

There may be some works that may be presented in a national congress of any society and you need to cite it in your research publication. This may also be a lecture delivered in any training program or summer course. The citation in the text is similar like any other research paper but when it comes to reference part of the paper, the reference includes not only name of the scientist with year of the congress held but also includes place of congress for easy identification as given below :

Islam, S.(1994). Occurrence of *Bivitellobilharzia nairi* in the captive Asian elephants (*Elephas maximus*) from Kaziranga National Park and Assam State Zoo, Guwahati.(Abstract). Sixth National Congress of Veterinary Parasitology, Jabalpur, p 2.

Chandra,R. On the occurrence of *Fasciola gigantica* infection in man in India. Seventh National (Indian) Congr. Parasitol., Raipur. Abstr. B-18 (1986).

Pathak, K.M.L. and Gaur, S.N.S. (1988). Surveillance of human taeniasis and porcine cysticercosis in certain parts of Uttar Pradesh. Proc. 2nd National (Indian) Congr. Vet. Parasitol., Bangalore. Abstr. p.5

Agrawal, M.C. (2002). A new strategy to diagnose animal schistosomiasis. In 7th *National Training Programme of ICAR on new approaches in diagnosis and control of Parasitic Diseases of domestic animals.* Veterinary College Bangalore, p 1-6.

FINAL VERSION

After writing the paper and correcting it several times, a version will emerge which is worth publishing in your view. Take this manuscript and consider in which journal you are willing to publish. After deciding the journal, take out a latest issue of the journal which has provided guidelines for contributors and you have to follow them strictly for giving final touch to your paper. Generally, there is no change in the basics of research publication in any scientific journal which is dealing your subject; Mostly the differences are in limitation of pages (papers are accepted generally with 5-6 pages, double space, 12 font, time new roman word file), photographs, graphs and their captions. There are directions for submission of photographs, graphs and tables, generally all are given on separate pages with captions.

The most important difference from journal to journal is about references and you have to follow the instructions accurately as per journal where paper is to be submitted. The differences may be at two levels- in the text or at last, list of references. You have to follow the manner prescribed in the journal while finalizing the manuscript and their pattern may be one of that as cited above. While submitting two or three copies (as per their demand) of the research paper to the journal, please keep one ditto copy for your record as the same will be required for checking corrections and further submissions. Many journals have started asking submission of the research papers via e-mail along with submission of hard copies by post; there are only a few journals in India, (though almost all foreign journals) which are totally dependent on submission of research papers on-line - these journals also do peer review on line and you should be computer savvy to handle computer system in this regards.

SHORT NOTE OR FULL RESEARCH PAPER

There are two ways of publishing the scientific work - one is Research note or short note and other is full scientific paper. The main criterion for it is importance of the work and whether sufficient data are generated for the paper. Short note does not reflect in any way inferiority of the work - it merely reflects the observations are at preliminary stage which require further investigations or it may be a contradictory observation against some existing theme. For example, we published a short note in

Current Science in 1985 (57) when we found excretion of *Schistosoma incognitum* eggs in albino mouse' feces where as all previous workers claimed otherwise ; or Dr SC Dutt published a short note in Nature when he observed development of both male and female *Schistosoma spindale* in guinea pigs which was hitherto considered a bad host restricting development of only male blood flukes. Many times it is the referee who asks the authors to change the whole paper into a short note after considering the subject and data presented in the paper. Generally short notes are accepted much faster and easy to publish than whole research papers.This is more so in important international journals.

SUBMISSION OF RESEARCH PAPERS IN ON-LINE JOURNALS

With the theme that scientific research should be open to access by any interested scientist from any part of the world, there is emergence of large number of online journals in almost all subjects from abroad as well as from India. The advantages of publishing in these journals are : it is peer reviewed, publish the article quickly (generally within two,three months) and copy right of the article rests with the authors. Nevertheless, it may be difficult to publish in an international online journal as they charge heavily for publishing the article in their journal-some times the charge reaches as high as $500/ for an article. Therefore, it is better to confirm publishing charges prior taking further actions.

There is no difference in preparation of the research paper for online journal and same procedures are followed as described above for any research journal. The difference comes while submitting the research paper- it is always through internet by logging and registering in the site of the journal. All the correspondence, submissions and corrections are made through internet without the need of any hard copy of any document.

REVIEW PAPERS

Senior scientists are invited to submit a review paper for a journal on the topic assigned by the editor- this topic is given looking to your experience on the subject though you may chose any other relevant topic to review. While writing the review article, it must be able to demonstrate the following :

- The main scientists and institutes, Indian and abroad, working in the field.
- Recent major advances and discoveries on the topic.

- Significant gaps in the research on the topic.
- Current debate or a critical view while discussing above
- Incorporating all the references,related to the topic.
- Your ideas how to tackle problems existing in the field.
- The article is written in past tense taking care of language and punctuation.

Its difficult to outline any other guideline for writing a review article. It is always better to read the previous review articles on the prescribed topic, prior writing your own article. As may be visualized the review papers are of two types. The first is on a general topic where all the past and present knowledge is incorporated and such review papers are of great value to the post graduate students or the beginners. The other type of review article is written on certain existing problem and deals the problem solely on different angles - development of drug resistance by *Haemoncus contortus* under Indian animal husbandry practices. I am referring some of my review articles whose reading will help in writing the review articles- these are of both types.

1. Agrawal M.C. and Shah, H.L. (1984). Stephanofilarial dermatitis in India. *Veterinary Research Communication*. 8: 93-102.
2. Agrawal, M.C. and Shah, H.L. (1989). A review on *Schistosoma incognitum* Chandler, 1926. *Helminthological Abstract* 58: 239-251.
3. Southgate, V.R. and Agrawal, M.C. (1990). Human schistosomiasis in India. *Parasitology Today* 6: 166-168.
4. Agrawal, M.C. and Alwar, V.S. (1992). Nasal schistosomiasis : A review. *Helminthological Abstract* 61: 373-384.
5. Agrawal, M.C. (1998), Parasitology in India since independence.. *Indian Journal of Animal Sciences*. 68 (8,special issue); 793-799.
6. Agrawal, M.C. (1999). Schistosomosis : an underestimated problem in animals in South Asia. *World Animal Review* 92: 55-57.
7. Agrawal, M.C. and Southgate, V.R. (2000); *Schistosoma spindale* and bovine schistosomiasis. *Journal of Veterinary Parasitology* 14: 95-107.
8. Agrawal, M.C. (2005). Present status of schistosomiasis in India. Proceedings of *National Academy of Sciences*, India. 75 (B), Special issue : 184-196.

9. Agrawal M.C. and Banerjee P.S. (2007). Problems confronting helminthic diseases of domestic animals in India. *Journal of Parasitic Diseases* 31: 3-13.

10. Agrawal M.C. and Rao V.G. (2011). Indian Schistosomes : A need for further investigations. *Journal of Parasitology Research* Vol 2011 doi : 10.1155/2011/250868.

CHAPTERS IN THE BOOK

However, the case is not similar while writing a chapter for a book. One must check whether the chapter is asked for a text book or a reference book. As is clear the text book for undergraduate students does not require dealing of the subject in details and disputed findings are avoided. In the reference book, it is the disputed results/observations which are discussed with explanations and lastly concluding present situation of the topic. However, the chapters, even in reference books, require some basic knowledge which may be included for understanding which is generally not required in the review articles.

1. Shah, H.L. and Agrawal. M.C. (1990). Schistosomiasis: in : *A review on Parasitic Zoonosis (*Ed. S.C. Pariza). ATIBS Publishers, Delhi 143-172.

2. Agrawal M.C., Shukla P. and Boregaonkar M. (1997). Preliminary screeing of some plants against fresh water snails. In Ecodevelopment and Environment (Ed. Singh, S.P., Singh R. and Ram Prakash) Vrinda Publications, M.G.Road, Jalgaon, India, Page 71-74.

3. Agrawal, M.C. Shah, H.L. and Chaudhary, R.K. (2003). Animal Parasitology: J.N.K.V.V. Research Achievements (1960-2002) (Ed V.P. Singh, M.S. Bhale, S.K.Jain and S.D. Upadhyaya) Faculty of Veterinary Science and Animal Husbandry, JNKVV, Jabalpur. 179-225.

4. Agrawal, M.C. (2003). Epidemiology of Fluke Infections. In *"Helminthology in India"*. (Ed M.L.Sood). International Book Distributor, Dehradun 511-542.

TECHNICAL REPORTS

It is always an honor to have a research scheme from any big, well organized funding agency as this does not only provide you chance to work on your line of research but also to enrich the department with some new and important equipments. Generally, in India, the funding agencies are government organizations like ICAR, ICMR, CSIR, DBT which sponsor large research schemes to eminent scientists.

These schemes are sanctioned generally for three to five years and every organization asks submission of annual report for intervening periods with final report on the last year. There are specific formats for submission of annual reports by each organization; however there are some common issues which are needed to submit in the annual reports.

The first page consists of general information like name of research scheme, sanction number, department where work is carried out, university code number, year of research scheme, principal investigator, co-investigators, total sanctioned amount, total expenditure incurred during the year (an important aspect is to submit annually AUC or audit utilization certificate duly signed by the competent authorities) and alike matters.

Some organizations ask to write short technical program of the research scheme while others ask only objectives of the scheme and work to be done in present and ensuing year. Then, you have to provide detailed work done during the year with presentation of data in different forms- including tables and graphs. Though, there is no need to apply statistical tests for each data, yet you should analyze the data at-least to preliminary levels like mean, range, percentage so that some conclusion may be drawn. As there is no separate part like discussion, you may cite some important references and critically analyze the results in light of previous works while presenting your results.

There are also columns like students guided for post graduate research work, seminars, conferences and workshops organized or attended and finally research papers submitted and published under the research scheme; while publishing research papers under the research scheme, always remember to acknowledge the help provided by the sponsoring organization; some organizations are asking attachment of research papers along with the report.

As mentioned in thesis sanction, it is always advantageous to keep a CD for the annual reports, submitted as this helps you in writing final report as well as in writing research papers.

18

Publishing Research Work

What so ever work you have done, it will not be recognized in the scientific world unless you publish it hence publishing the search work is of prime importance for a scientist. As it provides you recognition, you must pay greatest attention on quality of your research work which depends on the keenness of your observations. It is must to confirm and re-confirm your observations prior publishing them as once they are published, it will come in public domain with nothing you can do (except of publishing another paper to refute your earlier finding). As publishing the observations will form opinion about your scientific integrity and knowledge among the peers, the paper, if not cared and contained silly technical mistakes, will reflect ignorance of the workers on the subject. For instance, a paper was published in journal of veterinary parasitology on the porcine schistosomiasis from west Bengal where the pig schistosome was referred as *Schistosoma suis*, the name given long back by Rao and Ayyar in 1933 to the parasite but was characterized later as *Schistosoma incognitum* by Sinha and Srivastava in 1956 and is being used as *S.incognitum* since then thereby reflecting how casual are the workers in reporting their research work. Like wise *Schistosoma bovis* is an African schistosome, though suspected erroneously to be present in Indian continent, its presence is not possible in this region due to absence of the specific snail species. Even then, a paper on the parasites of Mithun *(Bos frontalis)* was published in recent times from North Eastern states of the country and Bhutan showing presence of *S.bovis* in these animals (98). As the workers have mentioned *S.bovis* in their research paper, it simply showed that they were completely ignorant about the existing scientific facts else they should have tried to confirm their findings by conducting further experiments. This casual treatment of the subject hinted to the scientific community that either the authors recorded *S.spindale* which is the schistosome of Indian continent and by ignorance described it as *S.bovis* or they have missed a good opportunity to describe a new schistosome species.

In contrast, we may cite the observations made by Dr SC Dutt while determining snail hosts for *Schistosoma nasale* which was, earlier, considered using both *Lymnaea luteola* and *Indoplanorbis exustus* ; meticulous observations led Dr Dutt to conclude that only *I.exustus* develops *S.nasale* (16) while *L.luteola* is refractory to this schistosome although it develops *Orientobilharzia dattai* whose cercariae closely resemble with those of *S.nasale* hence giving confusion. As these observations were made meticulously, the observation that *I.exustus* is the only host for *S.nasale* still holds true despite lapse of more than 50 years of the investigations. Here we may mention reporting of urinary schistosomiasis from village Gimvi of Ratnagiri district of India in 1952 by Gadgil and Shah (99), when all the scientific fraternity has ruled out possibility of existence of schistosomiasis in the country. This finding withstood all the tests because of two important observations by the researchers. The one was demonstration of schistosome eggs in human urine, and the other, most important, was that none of the sufferers have ever left India for any purpose. Therefore, the endemicity of the area for urinary schistosomiasis was never doubted though its intermediate host or even species of involved blood fluke have not remained beyond controversies. On the other hand,reporting of immunodiagnosis of Gimvi cases by ELISA in 1991, the most modern diagnostic method, (100) was shown being positive only in egg excreting human beings with its negativity in non-excreting human beings. The subsequent work on ELISA, though in other situations, have shown it not being 100% sensitive and specific thereby giving doubts about these results.

Thus, the long time impact of the research paper, as shown above, depends on its quality and how accurately observations are made ; the long term impact is free from journal of publication ; though for short term impact of the journal,where the paper is published, is important.

In India, presently, there are three journals which are devoted exclusively to Parasitology- journal of parasitology, journal of veterinary parasitology and Tropical Parasitology. The first is the official journal of Indian Society of Parasitology and covers all topics of Parasitology from Veterinary, Medical and zoology point of view. Presently, the journal is being published by Springer India Private Limited and accepts the manuscripts only on line thereby curtailing so many problems. Therefore, one may submit the paper on line by clicking for subscriber and may follow the subsequent developments.

Journal of Veterinary Parasitology, as the name suggests publishes the work related to veterinary site though it publishes papers dealing with parasitic zoonosis and parasites of wild animals also. The journal is presently published from the Division of Parasitology, Indian Veterinary Research Institute,Izatnagar, Bareilly and Dr PS Banerjee is Editor in chief of the journal. The journal still invites papers in hard copies in duplicate with all the photographs and tables.

Indian Academy of Tropical Parasitology is created in recent years at Department of Microbiology, JIPMER, Pondicherry with memberships mostly from medical fraternity. The Academy has started publishing Tropical Parasitology journal but only in soft copy with assurance to launch hard copy of the journal shortly. The Editor in chief, Dr SC Parija of the journal invites articles on line at his Pondicherry address.

Beside these three specific journals on Parasitology, there are two other important national journals which cover parasitology as well. The first is Indian Journal of Animal Sciences, being published from ICAR (Editor, Indian journal of animal sciences, publication department, krishi anusandhan bhavan, Pusa, New Delhi) and accepts papers on parasitology pertaining to domestic animals, wild animals and fishes. Though, its contents may be logged on by any one on its website, which is free, the journal still invites the papers in hardcopies with a CD of the research paper, including tables and figures.

Indian Veterinary Journal is one of the oldest scientific journal from India and is the official organ of Indian Veterinary Association having a permanent office at Chennai (Managing Editor, The Indian Veterinary Journal, No.11, Chamiers Road, Nandanam, Chennai-600035 India). This journal has started its publication since 1923 and continuing the same and is well recognized in Veterinary profession. However, its scientific impact is considered lowest (0.5) though this is the only journal which reaches to a large number of field veterinarians hence is of importance for publishing field oriented work.

Due to paucity of funds, all these journals are asking the contributors to be members of their association and asking charges for the tables, figures etc. In addition Indian veterinary journal has started asking Rs 50/ as handling charges for the paper hence its necessary to enclose this amount else the paper is not processed by the journal.

All these journals are important national journals and its good to publish the work in them. Please remember that almost all the national journals have stopped publishing repeat works without any new information. Therefore, either your work should deal with some new idea or if doing repeat work it should clearly be mentioned the reasons for the same and the theories which are to be re-tested along with new line of thinking.

Beside these journals, there are other important national journals which may include papers on parasitology provided the subject interest them. These are Indian Journal of Medical Research, (from editor, ICMR, Ansari Nagar, New Delhi), Current Science (Editor, Bangalore), Proceedings of NASI, Allahabad, Journal of ZSI. Almost all the Agriculture Universities, Pantnagar, Jabalpur, Mysore, Ludhiana, Marathwada etc are publishing their research journals which encompass Veterinary subjects including parasitology. As new veterinary universities are coming up in India, they have yet not started publishing their journals except TANUVAS, which is oldest among veterinary universities in India. Likewise, some research institutes, like IVRI, Hoffkin, AIITMH etc are publishing their own scientific journal. There are many other scientific associations, state veterinary associations which are publishing their own scientific research journal. Obviously, many of these journals are not regular in their publication and do no command any scientific impact.

PUBLISHING ABROAD

It is just a natural desire of the scientists to get published in foreign journals as it provides them international recognitions. But for publishing the papers in foreign journals, one must prepare the paper critically to minimize the chances of its rejection. When we are talking about foreign journals, it means international journals of repute having wide circulation else there are countries which are publishing international journals with limited circulation and limited impact in scientific community.

While searching international journals for submitting research paper,the first fact is to ascertain whether journal is charging per page for publishing research paper and you or your institute is ready to pay the amount that may turn out exorbitant when converted in Indian rupee (this is also true for online research journals). Indeed it is not only charges but matter of your research paper should synchronize to the policy of the journal. In Parasitology itself, there are many categories of the journals, giving emphasis on specific field of work. Thus, there is Systematic Parasitology, dealing with taxonomy, Journal of Helminthology (London) dealing exclusively to the helminthes, journal of Protozoology, exclusively on the protozoa, Journal of Experimental Parasitology accepting papers on experimental subjects, Annals of tropical medicine and Parasitology considers parasitic disease related matters like epidemiology, diagnosis, treatment etc. On the other hand, journal of Parasitology, International journal of Parasitology,, Parasitology today, consider all the subjects of Parasitology. There are also journals like Transactions of Tropical Medicine and Hygiene (London), American journal of tropical medicine and hygiene, South east Asian journal of tropical medicine and hygiene which covers parasitological topics, particularly disease related aspects and those which interest a larger scientific community.

You may submit full research papers as well as short research notes to any of these journals, looking to the subject of your paper. While submitting full paper to these reputed journals, please remember that a substantive work must be presented in the paper as mostly the papers are rejected due to less number of observations or using non-standard methods. Of course, the subject must be of international interest. For instance, the international scientific community may not be interested to know morphology, life cycle or chemotherapy against either much worked out parasitic species or one limited to a local geography. But they may be interested to know if your work is contradicting a well established hypothesis or observation or something new which has yet not been highlighted. By all means, its much easy to publish a short note in a foreign journal than full length article, specially when you wish to convey some thing unusual or more sought after by the scientific community. We may cite here one example. The prevalence of mammalian schistosomes in endemic areas is a widely recognized phenomenon but the endemic areas harbor generally one or two mammalian schistosome species ; contrarily, we reported (101) existence of five mammalian schistosome species in an endemic area (i.e. Jabalpur) hence it was interesting for international scientific community. Therefore, it was published as a short note by Transactions of Royal Society of Tropical Medicine and Hygiene, London. Likewise, cercarial dermatitis, caused by schistosome cercariae, is most common in endemic areas but there was no much reporting about the melody therefore bulletin of world health organization published this report from India in 2000 as a short note.

With the advent of computers and internet, it is heartening that all the international journals have become on-line journals hence authors may submit their papers directly through internet thereby shortening time of correspondence as well as other problems. Moreover, these journals generally communicate about acceptance or otherwise of the research papers within 2-3 months, thereby giving liberty to the authors to submit the papers elsewhere, afterwords.

ONLINE JOURNALS

Though there are now many on-line journals with free assess but almost all of them charge a fee in the name of processing fee ; therefore it is advisable to confirm about the payments either prior submitting the article or prior publishing them in these on-line journals. There is Journal of Parasitology Research and other is Journal of Tropical Medicine and Parasitology which are dealing the subjects and publishing the papers after charging processing fee.

19

Scientific Conferences

It is a must for the research scholars to attend scientific conferences of the Parasitology associations, after being members. It is advisable to become life-members of all three Parasitology associations, in India i.e. Indian Society of Parasitology, Lucknow, Indian Association for Advancement of Veterinary Parasitology, IVRI, Izatnagar and Indian Academy of Tropical Parasitology, JIPMER, Pondicherry. The annual conferences expose better chances to interlink with Parasitologists of the country and at times foreign scientists.

The associations dispatch the information about annual conference, much earlier, inviting abstracts of research papers which are to be presented in the conference. The invitation contains all the detailed information about hosting institute, city, duration of the conference (generally three days) with the provision of site visits, registration fee for members and guests, accommodation arrangement ; they also write on theme of the conference and technical seasons are divided in differe scientific seasons. For instance there may be scientific seasons like I. Morphology and taxonomy of Parasites II. Life-cycle, environment and epidemiology III. Pathology of parasites IV Diagnosis of parasitic diseases including immune-diagnosis V. Chemotherapy VI Control including biological control of parasites. You submit your abstract in the scientific stream where it is most suited and you wish to present it before the scientists of that stream. Please remember that only unpublished scientific work is to be submitted and presented in the conference provided it is not a lead paper which has different set of regulations (see below).

ABSTRACTS FOR CONFERENCES

Guide lines are given for submitting abstracts of the papers. It is with word limits (generally 200-400 words) starting with title of the paper, authors (underlining the person who is presenting the paper in the conference), institution address where work is done, and description of the work done. In short, the abstract starts with the present problem, how you tried to solve it,furnishing materials and methods, important results and significance of your results. There is no scope of giving references in the abstract (except one or two if extremely important) and you have to submit the abstract within stipulated time. Now, the associations have restricted number of abstracts being submitted which are published without peer review and distributed at the time of the conference. As an example,we are reproducing an abstract :

S-2:12 DEVOURING OF AMPHISTOME METACERCARIAE BY FRESH WATER SNAILS

M.C.AGRAWAL, Department of Parasitology, College of Veterinary Science and Animal Husbandry, Jabalpur -482001 (M.P)

While studying shedding behavior of amphistome cercariae from *Indoplanorbis exustus* snails, it was observed that the number of metacercariae attached to the glass wall or on green leaves, did not increase proportionately with advancement of days of cercariae shedding. A close scrutiny revealed that the snails were feeding on these metacercariae and passing the digested metacercariae i.e. empty cyst walls in their feces.

Further studies of the phenomenon showed devouring of the metacercariae by both the snails (*Indoplanorbis exustus, Lymnaea luteola*). When the snails consume metacercariae within 1-2 day of its formation, it was digested and only cyst walls were excreted with the feces. However, snails were unable to digest a week old metacercariae and passed them without digestion. It will be worthwhile to study the phenomenon in nature so that biological control of the amphistomes may be explored.

(abstract in 6th national congress of veterinary parasitology, college of veterinary science and A.H, Jabalpur, 22-24 October, 1994, page 14)

Some times, the association decide to publish whole paper, instead of abstract, and invite whole paper for publication in the proceedings of the conference. The research papers are prepared as is done for research journals and specific guide lines are provided for references etc to be followed by the contributor.

LEAD PAPER/THEME PAPER

Senior Parasitologists are asked to present a lead paper in the scientific conference mostly on the topic on which he is working since many years. The organizing secretary of the conference may ask either full lead paper or only abstract to be published. Instead of lead paper, you may be asked to submit a theme paper where topic of the theme is pre-determined.

These theme or lead papers are prepared like a review article (see above) where the senior Parasitologist high lights a particular problem, discussing past and present works and existing problems with some suggestions for future works. Obviously, the paper does not limit with your research work but includes those of others like any review paper.

1. Agrawal MC 2012 Why schistosomiasis should still be studied in Indian continent. Tropacon 2012. VIth National Conference of Indian Academy of Tropical Parasitology, 13 and 14 Oct 2012 Indore (Dr. SC Parija oration award lecture).

PRESENTATIONS IN CONFERENCE

Unlike to the abstracts, it's a different type of preparation for presenting the paper in the conference. Your paper is accepted either for oral presentation or as poster presentation. The scientific season will have a Chairman, Co-chairman and a Rapporteur. Depending on the number of oral presentations, chairman allow the contributor to present his paper to the audience within 5 to 10 minutes and afterwards, the audience is allowed to make questions about the work. The chairman prepares a summary of his scientific session, highlighting important observations in whole session and prepares his recommendations which are presented by each chairman during the valedictory function of the conference.

Though abstracts are prepared in running matter, presentation of the paper is done differently and more details are delivered. You have to describe, in short, aim of the work, materials and methods, results and conclusion- and you have to deliver the lecture within stipulated time. Earlier, this was done by preparing micro -slides, to be shown by a slide projector, later the matter was taken on transparencies of A4 size to be exhibited through a projector. In present time, with advent of computers, presentation has been made much more easy and more interesting with less cost. Now presentation is prepared on "POWER POINT" where not only writing is easy but you may insert any photograph, graph or one two important reference with out any problem. This presentation is copied on a CD or taken in a pen drive.

As time is limited for presentation, its advisable to prepare 6-8 slides avoiding too much material on a page as this poses problems of visibility. The first slide should contain title of the paper, name of all the workers, underlining the presenter and address of the institutions where work has been carried out. The second slide may contain present problem while third may contain your materials and methods. You may mention the reference if your method is standard else may provide more details. Fifth, and sixth slide contain important results while subsequent slide shows conclusion of the work. Its better to depict the results with graphs and photographs as only numbers are difficult to appreciate by the audience. The power points are only for your help, and visualization by the audience and you should avoid to read the presentation as it reflects badly. Of course, last slide may contain thanks, showing your gesture as well as end of your presentation.

POSTER PRESENTATION

If your paper is accepted as poster, you have to prepare the poster following all the instructions; important is the size of the poster where you have to depict your presentation. Here, specified space is provided and you write details of materials and methods, results and conclusion - include original photographs, line drawings and graphs which facilitate explanations. It's a bad advice to copy some photographs either from a text book or from some other research paper except in discussion, if necessary. Students are advised to prepare all the presentations their own which will develop confidence but after taking help from their guide, seniors. Again, it will be more appropriate to conduct rehearsal in the department which gives a chance to find possible questions and also about your preparations.

Generally poster presentation is asked to young parasitologists or post graduate students who wish to exhibit their work before a scientific gathering. The main difference from oral presentation is that the posters are depicted at certain places where poster presenter should remain there to explain any point to the visitors. A time is allotted where scientists visit the posters and evince interest raising questions which are answered by the poster presenters.

AWARDS

For encouraging the research scholars, every conference have the awards for best presentations- both oral and poster presentation. Naturally, it attracts attention if you win any of these awards in the conference.

LEAD PAPERS

Presentation of lead paper is quite different then presenting an abstract before the conference. Now, there are all the facilities to prepare the lead papers also on the power point of a computer. However, its headings are different than that of abstracts. Here your presentation is about the past and present works carried out by the scientists, high lighting short falls of previous researchers, how you tried to solve the problem, your successes in some areas and present problems. Its better while expressing present problems some ideas to mention how to solve the problem.

It is advisable to include original matter for presentation and avoid to use text book presentations - this gives a bad idea about your authority on the subject unless you are to prove a opinion or otherwise. At the most you may exhibit photographs, graphs of other works also but by giving due credit to them. A time (20-30 minutes) is also allotted to present the lead paper which is followed by a question session.

20

Extension Literature

As the Veterinary colleges, and Parasitology being an integral part of them, are part of Agricultural universities framed on USA land grant pattern, or separated as Veterinary university imbibing the same philosophy, extension is an integral part of the system. Here, a team of the experts visit the villages to teach the villagers about the basic problems, faced by them in day to day dealing with animals. For explaining a particular problem to the villagers, the experts generally take help of the pamphlets which they have prepared. These pamphlets or extension literature deal the subject with common sense and illustrate the material with line drawings.

SELECTING THE TOPIC

Whenever you are publishing these pamphlets, select the topic in the manner most beneficial to the animal owners; you are not making them expert of the subject but training them some basic facts which can increase animal production or minimize animal mortality. For instance, instead of publishing pamphlets separately for fasciolosis, amphistomosis, schistosomosis, it is better to publish only one pamphlet informing the animal owners about the important fluke infections, how they affect their animals, most important symptoms, and how these animals can be protected from the infection; the whole purpose of the pamphlet is to protect their animals from fluke infections and if it happens to indentify it in their animals (as treatment of the animal is bound with veterinary practices acts, you can not recommend treatment of the animals for the ailments). There are many topics on which pamphlets have been prepared by the scientists and faithfully distributed in the villages. These pamphlets are written in local languages and published generally restricting to one or two pages.

The target audience is the villagers or animal owners, many of them may be primary pass while a few of them may have passed high school. Therefore, you have to prepare the pamphlet in such a fashion that it may be understood by these semiliterate persons. You can not use tough words which are difficult to understand, neither many semi technical words that may complicate meaning of the subject. It will be an absurd idea to use any reference in these extension pamphlets. Here illustrations or some real photographs are of great help as these are easily grasped by any person. It is advantageous to insert procedure,approximate cost and contact details where a farmer may like to sale his animal produce. If there is any plan of the Government of providing subsidy for particular project, its details should invariably be furnished. While preparing the extension material, always insert full address and contact details of the persons whose services may be taken up if needed.

PAMPHLETS IN PUBLIC HEALTH

Above pattern has now been followed even by medical team while dealing with infectious diseases or public health problems. Thus the PHC or experts of the subject are educating the villagers in the endemic areas through extension pamphlets about the benefits of use of insecticide impregnated mosquito nets in place of simple net to get rid of mosquito population. There are pamphlets distributed in affected areas- how iodized salt can protect them from Goiter or how they can be protected from "*Hathi pau*" or filariasis. While preparing the pamphlets all the above details are included with addition of name of health centre where suspected persons may visit for disease verification.

Now on the same pattern as followed for villages, many hospitals and medical institutes have started publishing extension material for distribution among their patients (where patients are highly educated). Even where a medical team is investigating parasitic infection like amoebiasis or hookworm infection, it circulates the pamphlets informing general semi-literate population in most non -technical or semi-technical way how the disease spreads among population, symptoms seen at various stages, how man can avoid such infections and when and where they can consult the physician.

NEWS PAPER ADVERTISEMENTS

Two extension methods are being followed for covering targeted population. One is publishing and distributing pamphlets, as mentioned above, and other is advertisements in the papers highlighting related facts.

The advantage of inserting advertisement in news papers is that the experts have not to visit the targeted population but his message is conveyed more effectively and more widely. Obviously, the advertisement could be read and understood only by more literate persons thus excluding the other group of persons. Secondly, advertisement costs much more than the pamphlets, hence restricted matter is inserted to curtail the expenditure; again, every time expenditure will be incurred while inserting the advertisement which is not the case in publishing the pamphlets.

Though matter is restricted due to limited space, a better language is used for conveying the message owing to change in readership.

USE OF INTERNET

With the invent of computer, lap-tops, and internet, the village Panchayat has now access to any source of information as many of them are connected through internet. To cope up with this new development, scientific community has come forward to prepare different presentations, videos and other forms to convey the message in more simpler form ; need of using words has been minimized and more and more visuals are used for conveying the message ;it may be about A.I. or forming vermin-compose, or how a disease is spreading in the animals, information on the projects where farmers are entitled for getting subsidy.

The greatest advantage of generating internet transmitting information is that they are very cheap, can be accessed by a large gathering from any where of the geography and can be used multiple times without incurring any extra cost except that of internet connection. Therefore, this mode of communication is becoming popular for conveying the message as well as for contacting the desired persons as and when needed.

SEMI TECHNICAL PUBLICATIONS

There are many magazines which are catering to general public and are not publishing scientific articles but dealing the subject in semi scientific way so that it may be understood by an educated person who is not a specialist of any subject. Some magazines have local circulation while others, national circulation though its always advantageous to publish in national magazines. The magazines are published either in English, Hindi or in local languages. Some important magazines are ; Indian farming (ICAR), Kheti, Livestock advisor, Poultry advisor, Blue cross book, the veterinarian, Dairy farming. As the name suggests they are dealing different subjects and non is dealing any specific subject (e.g. Parasitology or microbiology) but a group of subjects like poultry or dairy while "the veterinarian" is more technical magazine, made for field veterinarians. Most of these have readership among the persons who wish to understand the subject in non-technical way. In general, the readers are educated farmers, animal owners, veterinarians, science graduates, personnel dealing with the livestock industry. At times, our policy makers also consult these magazines as they are easy to understand.

The aim of publishing articles in such magazines is different than that of publishing pamphlets. The readers are more educated and wish to acquaint themselves not only with knowledge but latest knowledge, in semi technical language, that has emerged in recent years. Thus in many ways, these semi technical articles are like review articles; the difference is only in readership and type of presentation. In technical review articles, the readers are professional of that subject which is not the case with such magazines. Therefore, it is preferred to present the subject with minimum use of technical words and to illustrate the subject with photographs and line drawings. Initially, author of the article explains basics of the topic in simple language and then expose the reader with new developments taking place in the subject. If the magazine is for general public, there is no need of inserting references but this is done while contributing articles for Blue cross book or the Veterinarian.

21

Internet Mail, Blog, Website, You Tube

Till twentieth century information technology was restricted to European and American countries but entering in twenty first century witnessed passage of this technology with a vast speed in all other countries making a spectrum of changes in communication and transmission of knowledge that has altogether changed scenario of whole world. Earlier communication to international scientists was costly, time consuming while now its only a click away. So is of transmitting reprints of research papers - earlier they were dispatched by mail, now these reprints and other matters are available in soft copy and are attached with e-mail that is transmitted instantly.

INTERNET CONNECTION

When you are in research profession, it is important to have link with other scientists and scientific community. The basic requirement for this link is a desk top computer or a lap-top or a smart phone (use of smart phone alone will create problems in typing and transmitting larger size of text which is easily performed on lap-top or desk top) and you have to connect this gadget with any internet facility - it may be either through BSNL or any private dealer like Idea, Air tel etc. Once you are connected through internet, there are various options of communication which may be followed. Some of them are:

E-MAIL ACCOUNT

This is the basic requirement for sending or receiving mail from any person within country or outside the country.Therefore,it is important to open e-mail account which is basic for opening new blog or any other website. You may open email by logging www.mail.com/mail/create-email-account/

You will be given a user name and a password for your e-mail account which are remembered and entered always whenever you are opening your e-mail account. Interestingly, you can operate your e-mail account from any part of the world.

This e-mail system is not only good for communication but you may attach number of files and send them to the desired persons. Additionally, you may move any important documents to specific topics by creating such topics - This place is safe for preserving important files with no chance of destroying by virus or computer default. However, please remember that every thing which you are posting on e-mail is accessible and may be traced by interested persons.

FORMATION OF BLOG

I think the second important site in present times for a scientist is to open a blog - again you will be provided a unique name of the blog by which any other person may log in to your blog and a password which is a secret number by which you may edit and add any matter in the blog; prior that you should have your email account. You may open the blog by logging following site - https://accounts.google.com/signup?service = blogger.

The advantage of the blog is that it is free of cost and you can add any typed matter of considerable length along with photographs which is not possible in other operations. These blogs are used to convey, message, posting research information and any other material. These blogs are from individuals as well as from organizations e.g. journals, corporations, science institutes and gives an opportunity to link with any of them to receive latest information from that scientist/organization. However, it must be remembered that when you are in service, do not post any critical material as disciplinary action may be initiated against you on the basis of your blog posting. It must be very clear what you wish to accomplish through this blog; if you wish to establish yourself in particular field, it is better to invite comments on your matter by posing some questions and respond to the comments.

The blog can be logged from any where of the geography but to increase viewership, it is better to link your blog with other blogs so that a better exchange could occur among a spectrum of scientists. You may identify science blogs from following sites :

- Yahoo science blogs
- Bloglines feed directory
- Google blog search

- Interestingly you may open more than one blog with different name and purpose. I am mentioning my three blogs showing importance of each one
- www.indianparasitologists.blogspot.com
- www.indianschistosomiasis.blogspot.com
- www.JVCAlumniAssociation.blogspot.com

Obviously no blog can be created in future with similar URL and you have to make some changes while opening a future blog on related topic.

WEBSITES

The same operations which are performed in blogs may be carried out by creating a website. This website enables you to up load any information and may be divided in different sections,visible on home page. The greatest advantage of the website is that it preserves the headings and sub-headings as such while the blogs store the old postings by the dates hence one can not ascertain the subjects dealt in those postings. The disadvantages of the website are that they are chargeable annually (though there are free sites also) and if not operated frequently, there are all chances of hacking the website. Moreover, sufficient matter should be added continuously to maintain interest of the viewers in the website, which may not be possible for an individual scientist. As an example, the department of Parasitology, College of Veterinary Science, Jabalpur has created a website www.drscdutt.com and uploaded material related to DR SC Dutt, his memorial lectures etc ; however we fail to load new matter frequently hence the website could not generate much interest to the viewers and was frequently hacked. Therefore, it is better for a young scientist to start with the blogs and when he has a standing he may create his own website.

LINKED IN

This is a new portal where the professionals are loading their bio data and other details particularly specializations in different fields. You may open a free account in the Linkedin by logging https://www.linkedin.com/reg/join

This is the site where you can contact various professionals of your field or related field. The portal provide to communicate free of cost to important professionals but when you desire to contact individually for getting a job or any other assignment, the portal asks registration by paying a fixed amount.

PROFESSIONAL GROUPS

There are many professional groups which are operating their portals and they are free of cost. Once your e-mail is in operation, you will receive messages to join these professional sites. There are sites like Helminthology group; Parasitology group etc which are of interest to us. These sites open discussion on some important topic, you may participate in such discussions or open your new discussions; some research scholars are loading their important queries or requests and there are scientists who are attending them.

YOU TUBE

This video sharing website was founded on 14th February, 2005 and is headquartered at California, USA. It is a most popular site (http://www.youtube.com) for watching films and viewing songs. Interestingly, there are many other materials that may be searched on you-tube. For example, there are 200 videos on animal husbandry; all the details in different videos regarding Gobar-Gas Plants or Vermin composting methods; even you may search under control of fluke infections where you can observe management tools for control of liver fluke or *Paragonimus* and dissection of liver fluke infected snail.

Some of the material in above videos may not be fitted in your local conditions. In that case, you can register yourself on this site (www.youtube.com/playlist?) and upload your own prepared video on any scientific event- it may be related to delivering your lecture in a scientific conference or a video related to control for fluke infections in ruminants etc. As the matter is visual, this can easily be grasped even by a non-literate person, a considerable population of animal owners. In other words you-tube has replaced the need of distributing pamphlets in the villages ; now you can up-load the videos on you-tube which may be played by any one having internet connection.

FACE–BOOK

This is presently most popular website which has been a good source to link with your friends and any other person. Now even organizations have opened their account and you may join them to communicate. For opening a face book account (www.facebook.com/pages/creat-new-account../) you should have a Gmail account. This website enables you to post photographs and write ups with the choice whether these should be viewed by selected group of persons or it is for public. Even you may block viewing your face book by a particular person who by mistake was selected as friend.

The world is changing very fast and every moment some new gadget is coming for the help of man communicating with each other from any part of the world. I have not mentioned another popular site i.e. twitter as it is mainly for media persons who wish to communicate only in few lines and almost daily.

Appendix

NOTES AND REFERENCES

Here, we are mentioning notes, websites and references which may be consulted for getting further details on the topics mentioned in the text:

PART A

1. Vikram Doctor (2011). Agricultural research's real estate problems. The Economic Times, 14th Jan, 2011, New Delhi (there was a series of articles in daily news paper economic times on such topics, at that time).
2. Agrawal MC (2012). Schistosomes and schistosomiasis in South Asia. Springer India Pvt Ltd, New Delhi. page 351 (this is the first book on schistosomiasis in South Asia, dealing all aspects of the infection).
3. Shri Ram Charit Manas was completed within 2 years, 7 months 26 days in the samvat year 1633 on "Ram Vivah day" by shrimad Goswami Tulsi Das jee. The text (written in Avadhi- a form of Hindi) has been published by Geeta Press, Gorakhpur and interpreted in Hindi by Shri Hanuman Prasad Poddar. Till the samvat 2045, there have been 69 editions of the book by Geeta Press, Gorakhpur, UP.
4. The meaning of Ayurveda is Ayus=life, Ved=knowledge. It is a part of Atharvaveda which is written about 1000 years BC (according to some veda are written about 5000 years back) and an important part of vedic life of the Indians. There are certain principles laid down for long human life. Dhanvantri, who is considered incarnation of lord Vishnu, is considered the Great Physician, and writer of Ayurveda.
5. Sage Maharshi Ved Vyas has written 'Maha Bharata" (in Sanskrit) and "Shrimad Bhagvat Geeta" is a part of this epic. Its period of writing is about 3500 years back. Geeta Press, Gorakhpur, UP has

published 34th edition (with Hindi translation) till samvat 2045. This is a sacred document of Hindu religion and has been interpreted by many scholars including Sarvapali Radha Krishnan, Mahatma Gandhi and Vinoba Bhave etc.

6. History of agriculture domestication of plants and animals has developed around 12,000 years ago (en.wikipedia.org/wiki/History_of_agriculture). In Indian subcontinent (about 9000 BCE) wheat, barley, sheep and goat domestication was followed. Also see Hillman GC 1996. Late Pleistocene changes in wild plant-foods available to hunter gatherers of the northern Fertile Crescent: Possible preludes to cereal cultivation. In D.R Harris (ed) *The origin and spread of Agriculture and Pastoralism in Eurasia.* UCL BOOKS, London pp 159-203.
7. Agriculture resulted in labor divisions and development of cities and human civilization. Some well developed settlements had arised in Mesopotamia, Egypt, India, China on major rivers of these countries. Also consult - Allchin Raymond (editor) 1995. The Archaeology of Early Historic South Asia: The Emergence of Cities and States. Cambridge University Press, New York.
8. Bhagwan Das (1976). Carak in V. Raghavan ed. *Cultural leaders of India; Scientists* Publication division, Ministry of Information and Broadcasting, Government of India, New Delhi (the book provides authentic accounts of the life and work of great figures like Dhanvantari, Caraka, Susruta, Arya Bhata etc who influenced mind and life of man kind- it also briefly tells history of ancient science in India).
9. History of small pox ((en.wikipedia.org/wiki/History_of_smallpox). Its emergence in human population dates back 10,000 BC. The earliest credible evidence seen in Egyptian mummies. Also consult - Erratum in 1998. Smallpox : the triumph over the most terrible of the ministers of death. Annals of International Medicine 128 : 787 (pubmed/9341063) May also consult WHO small pox and its eradication. Public health No.6. ISBN 92-4-156110-6 for more details.
10. Stefan Riedel (2005) has written in detail about Edward Jenner and the history of smallpox and vaccination in the Proceedings of Baylor Univesity Medical Center Jan 2005 issue (vol 18, page 21-25) and may be logged www.ncbi.nlm.nih.gov/pmc/articles/PMC1200696/
11. Prior germ theory of Louis Pasteur (en.wikipedia.org/wiki/Germ_theory _of_disease) or at the time of Edward Jenner, Miasma

theory was existing which suggested it is pollution by bad air, responsible for spread of diseases (also see note 39).

12. Charles Darwin (1859). The Origin of species, first edition published on 24th Nov 1859 in England. (but first Indian edition was published in 1992 by W.R.Goyal Publishers and Distributors, Delhi and reprinted in 2006).

13. These are the Antibodies (called Plantibody) produced by genetically engineered plants. Also consult-link.springer.com/reference workentry/10.1007%2F978-1-4020-6754-9_12950

14. Abdul Kalam APJ (2003). Ignited Minds. Penguin Books India. Page 224 (ISBN 9780143029823). The former President of India and widely acclaimed Bharat Ratna personality has published a number of books in association with Penguin Books India and this is one of them.

15. Sen Amartya (2005). The Argumentative Indian (writings on Indian Culture, History and Identity). Penguin Books Ltd, London, England (printed in India by Gopsons Papers Ltd, Noida-ISBN014101211-0).

16. Srivastava HD and Dutt SC (1962). Studies on *Schistosoma indicum.* Research Series bulletin No. 34. Indian Council of Agricultural Research New Delhi (the monograph deals with the studies related to *S.indicum* with the techniques applied in studying the parasite).

17. Agrawal MC (1999). New methods of maintenance of fresh water snails for schistosome infections in the laboratory. *Indian Journal of Animal Sciences* 69: 301-303.

18. Dutt SC (1957). Development of female *Schistosoma spindale* in guinea pig. Nature 179:1359 (this is the short note challenging the hypothesis that guinea pigs possess some "host factors" which prevent development of female *S. spindale* ; their identification may help in control of schistosomiasis : (Full paper was published in 1962 in *Parasitology* 52: 199-206).

19. You can read the whole nobelprize lecture by logging www.nobelprize.org/nobel_prizes/medicine/laureates/1902/ross-lecture.pdf

20. There are various publications on Einstein. You will come across some of them by logging www.arxiv.org/ftp/arxiv/papers/1205/1205.5539pdf

21. May get biographical details of Srinivasa Aiyangar Ramanujan with many references by logging www.history.mcs.st-and.ac.uk/Biographies/Ramanujan.html

22. This site (www.tagoreweb.in/pages/RTagore.aspx) provides in English a short biography of Tagore and that at the age of 17th he was sent to England for formal schooling but did not finish there. The other pages led us to inform, in Bengali, about his various songs, plays, novels etc.

23. The London School of Hygiene and Tropical Medicine (www.lshtm.ac.uk/library/archives/ross/biography/) has run a project on Ronald Ross and has provided many details about this nobel laureate who became Professor in tropical medicine at Liverpool School of Tropical Medicine and Director of the Ross Institute and Hospital for Tropical Diseases till his death in 1932.

24. You will get all the details of Steve Jobs, his birth, education, occupation etc (www.biography.com/people/steve-jobs-9354805) and also of Isaac Newton and Leonardo da vinci

25. There are many writings on Bill Gate, the American business-magnate, philanthropist,inventor and computer programmer. May check his early life, about Microsoft by logging en.wikipedia.org/wiki/Bill_Gates.

26. There are publications in different languages on Swami Vivekananda. This site (www.vivekananda.net/PDFBooks/lifeofSVVol2.pdf) is created by the Pegents of University of Minnisota under Himalayan Series and covers the Swami in Himalayan series No. XXVII.

27. National institute of parasitology (the issue was touched while delivering DR SC Parija oration award at Indore. A more detailed account was published by PARA SIGHT volume 2 of IAAVP and is also available on my blog www.indianschistosomiasis.blogspot.com)

28. This book "In Exile" written by Ronald Ross was reprinted by Malaria Research Centre, New Delhi on the occasion of second global meet on parasitic diseases held on 20th August, 1997 at Hyderabad. This book contains poems, penned by Ross and preface to the third edition of the book, written in 1931.

29. CHAUHAN, CPS (2010). Higher education in India: Issues and challenges. In Khurana SMP and Singhal PK, editors. *Higher education, quality and management* Gyan Publishing House, New Delhi, p 75-89.

30. Anonymous 2010. Ras Rang, Cover story. Dainik Bhaskar daily news paper Jabalpur Dated 3rd October, 2010.

PART B

31. There are many articles on history of parasitology, depending on geography, type of parasite, host. You may refer en.wikipedia.org/wiki/Parasitology where all branches of Parasitology are referred. For history of human parasitology, you may log www.ncbi.nlm.nih.gov/pmc/articles/PMC126866/

32. The library of ancient texts online (https://sites.google.com/site/ancienttexts/gk-a3 refer Greek texts as well as sites where translations of the text are available. The history of animals (en.wikipedia.org/wiki/history_of_animals) refers Aristotle's text "Historia Animalium".

33. In his paper (www.sju.edu/int/academ) Karen R Zwier has discussed "Aristotle on Spontaneous Generation".

34. The science museum of London explains about Miasma theory and provides other related references as well (www.sciencemuseum.org.uk/broughttolife/techniques/miasmatheory.aspx).

35. This site (www.history_of_the_microscope.org/anton-van-leeuwenhoek-microscope-history-php) provides the photograph of the microscope developed by Leeuwenhoek and his discovery of single celled organism, whose existence previously was unknown.

36. You may find life and work of Andry by logging (www.ncbi.nlm.nih.gov/pmc/articles/PMC2908340/) and that his work lacked experimental proof but depended on microscopic observations by logging (en.wikipedia.org/wiki/Nicolas_Andry)

37. The encyclopaedia Britannica (www.britannica.com/EBchecked/topic/55517/Agostino-Bassi) refers his work on silk worm disease and his generalization that many diseases of plants,animals and man are caused by animal and vegetable parasites.

38. Details of life of Ignaz Semmelweis and his experiments on hand washing are given on this site (en.wikipedia.org/wiki/Ignaz_semmelweis)

39. This site (www.pasteurbrewing.com) provides many detailed works of Louis Pasteur including Pasteurization, spontaneous generation and lead to article on germ theory of disease. For reading English version of lecture on germ theory to the aetiology of certain common diseases,delivered by Louis Pasteur before French Academy of Science on 3rd May, 1880, you may log https://ebooks.adelaide.edu.au/pasteur/louis/exgerm/complete.html

40. This site (www.life.umd.edu/classroom/bsci424/BSCI223Website Files/KocksPostulates.html) informs about four important Koch's postulates and also lectures on bacteriology. The site is prepared by University of Maryland.

41. Its better to read the novel lecture of Alphonse Laveran delivered on 11th Dec 1907 on Protozoa as cause of diseases by logging www.nobelprize.org/nobel_prizes/medicine/laureates/1907/laveran-lecture.html

42. Paleoparasitology has provided evidence of hookworm eggs in human coprolites (fossil fecal material) in south American archaeological digs dating back 7,200 years (micro.magnet.fsu.edu/optics/olympusmicd/galleries/oblique/brazilianhookworm.html)

43. You may search Biography, nobel lecture and Banquet speech of Paul Muller by following www.nobelprize.org/nobel_prizes/medicine/laureates/1948/muller-facts.html

44. Chowdhury N (1994). Helminths of domesticated animals in Indian subcontinent. In Chowdhury N and Tada I, editors, *Helminthology*. Narosa Publishing House, New Delhi page 73-120 (this important article reviews important helminthic work carried out in India and accompanied countries).

45. Bhatia BB (1997). History of parasitology with special reference to Veterinary Parasitology in India. *Journal of Veterinary Parasitology* 11:111-124 (this paper describes chronologically developments in parasitology without much debate).

46. Agrawal MC (1998). Parasitology in India since independence. *Indian Journal of Animal Sciences* (Special issue), August 1998 : 793-799. (this paper, with out references, discuss important parasitic infections taken up by the Indian Parasitologists, mainly related to Veterinary Parasitology).

47. You can read about Kuchenmeister experiments and other details in the ebook "*Taenia solium* cysticercosis from basic to clinical science" edited by Gagandeep Singh and Sudesh Prabhakar by logging books.google.co.in/books?id=ICAd-uTQC&pg=PA160&dq=Freidrich+kuchenmeister

48. The book on P.C.Abildgaard (1740-1801) biography and bibliography by Siguard Andersen (https://books.google.co.in/books?id=cQVVAAAAYAAJ&q=parasitology&dp=ABILGAARD..) also provides his experiments with *Lingula*.

49. To study genomics in parasites, you may log to https:books.google.co.in/books?isbn=142000870 and also on parasitic nematodes vide ISBN 1845937597

50. Hsu HF and Hsu SYLi (1956). On the infectivity of the Formosan strain of *Schistosoma japonicum* in *Homo sapiens*. American Journal of Tropical medicine and hygiene 5 : 521-528 (also see their paper) "The infectivity of four geographical strains of *Schistosoma japonicum* in the rhesus monkeys" Published in *Journal of Parasitology*, 1960, vol 46 Page 228.

51. For basic knowledge on *Bacillus thuringiensis* pl see (en.wikipedia.org/wiki/ Bacillus _ thuringiensis). For its effect on mosquito larvae consult paper of Poopathi Subbiah and Abidha published in *Journal of Physiology and Pathophysiology*, year 2010, vol 1, Page 22-38.

52. This is the study of parasites from the past and may have general knowledge by logging en.wikipedia.org/wiki/Paleoparasitology.

53. This site (www.iloveindia.com/history/ancient-india/maurya-dynasty/chanakya.html) deals with history of India mentioning great figures including Chankya who is also known as Vishnugupta.

54. There is brief about Satyendra Bose and his association with Albert Einstein and about his theory on god particles (www.biography.com/people/satyendra-nath-bose-20965455)

55. The book "Remembering Dr SC Dutt: The Parasitologist" was published in 2002 by the department of parasitology, college of veterinary science & AH, Jabalpur (a revised edition of the book is published by Har-Anand Publishers, New Delhi, 2015). The article "metamorphosis of a student" is also part of the book and is on the blog www.indianparasitologists. blogspot.com

56. Sinha PK and Srivastava HD 1965. Studies on *Schistosoma incognitum*, Chandler, 1926 III On the host specificity of blood-fluke. *Indian Veterinary Journal* 42: 335-341

57. Agrawal MC, Panesar N, Das M (1985). Patent infection with *Schistosoma incognitum* in white mouse (*Mus musculus*). *Current Science* 54: 640-641

58. Montgomery RE (1906). Observations on Bilharziasis among animals in India. *Journal of Tropical Veterinary Science* 1(15-46): 138-174 (in fact Montgomery has published two papers in same volume of the research journal).

59. Agrawal MC (1974). Studies on the heterologous immunity in schistosomiasis. Ph.D thesis, Jawaharlal Nehru Krishi Vishwa Vidyalaya, Jabalpur, MP.

60. Agrawal MC (1998). New techniques for infection and recovery of schistosomes from animals. *Indian Journal of Animal Sciences*. 68: 521-523.

61. There is doubt about the story of golden crown in scientific fraternity. For more details on this Greek scientist, please refer (en.wikipedia.org/ wiki/ Archimedes).

62. There are several sites on Alexander Fleming. For preliminary information may see (education-portal.com/ academy / lesson/ Alexander-fleming-discovery-contributions-facts.html). For more details about accidental discovery please log (en.wikipedia.org/ wiki/ Alexander_Fleming).

63. This site informs about Mendel's laws (anthro.palomar.edu/ mendel/ mendel_ html).

64. Norman Emest Borlaug is credited with developing a dwarf wheat variety which is responsible for green revolution. The scientist was awarded noble peace prize in 1970 for his pioneer work which saved billions of lives (en.wikipedia.org/ wiki/Norman_Borlaug).

65. World over, vaccine development against malaria is carried out but no commercial production is done till now. Better to log this site of WHO (www.who.int/ malaria/ areas/ vaccine/ en/) which gives a general view about malaria vaccine.

66. The nobel prize was instituted as per will of 27th Nov, 1895 of Alfred Nobel for outstanding achievements from all corners of the globe in Physics, Chemistry, Medicine, Literature and for work in Peace (later, more subjects have been added) see www.nobelprize.org/ alfred_nobel/.

67. Nagavara Ramarao Narayana Murthy started Infosys, which is now a multinational company, with an initial capital of Rs 10,000/ only ; may get details on the site en.wikipedia.org/ wiki/ N_R._ Narayana_ Murthy.

68. This site informs about history of Ranbaxy and work of Bhai Mohan Singh (www.maheshsundar.com/ home/ maheshsundarcom – pharma-updates/ the_history_of _ranbaxy) and wikipedia for initial information.

69. The details of Dr Reddy are provided in the website of the company; may reach the site by logging (www.drreddys.com/aboutus/our-founder.html)

70. The details with more references may be seen in en.wikipedia.org/wiki/Kiran_Mazumdar-Shaw

71. This company is established to grow and convert seaweeds into biofuel (www.sea6energy.com)

72. The Stempeutics Research Private Limited, Bangalore is working on stem cell based medicinal products (www.stempeutics.com)

73. Dr Raghunath Mashelkar, former director general of CSIR is co-founder of invictus oncology. Details of the company may be seen on their site - www.invictusoncology.com

74. This Padmashree awardee is considered father of Indian poultry industry who could establish Venkateshwara Hatcheries Ltd, Pune by sheer hard work and interest. This site is important to read initial life of Dr B V Rao (www.livelihoods.net.in/sites/default/files/pdf/vasudev%20Rao.pdf)

PART C

75. The on line public access catlog was started from (1975) by Ohio state university and you get all basic knowledge on OPAC by logging en.wikipedia.org/wiki/Online_public_access_catalog

76. Soulsby EJL (1982). Helminths, Arthropods and Protozoa of domesticated animals. 7th edition, Bailliere Tindall, London (a lower cost edition of ELBS was available for commonwealth Countries ; as per my knowledge, no new edition has been published of the book since 7th edition)

77. Silverman PH (1965). *In vitro* cultivation procedures for parasitic helminths. In Ben Dawes edit. *Advances in Parasitology* Volume 3, Page 159-222.

78. Silverman PH and Hansen EL (1971). *In vitro* cultivation procedures for parasitic helminths : Recent advances. In Ben Dawes edit. *Advances in Parasitology* Volume 9, Page 227-258.

79. Gibson TE (1964). Recent advances in the anthelmintic treatment of the domestic animals. In Ben Dawes edit. *Advances in Parasitology* Volume 2, Page 221-257.

80. Keeling JED (1968). The chemotherapy of cestodes infections in Goldin A and Hawking F editors. *Advances in chemotherapy*. Volume

3, page 109-152 (also see same volume page 153-251 for chemotherapy of trematode infections ; same volume page 39-108 for chemotherapy of trichomoniasis).

81. Alwar VS and Lalitha CM (1961). A check list of helminth parasites in the department of Parasitology, Madras Veterinary College. *Indian Veterinary Journal* 38 : 142-148 (this is addition then published by Dr VS Alwar in 1954).

82. Srivastava HD (1945) A survey of the incidence of helminth infection in India at the Imperial Veterinary Research Institute, Izatnagar. *Indian Journal of Veterinary Science* 15 :146-148 (in the Presidential address in 1971 of 41st Annual session of Indian National Academy Sciences (Biological sciences) HD Srivastava further discussed on helminth parasites of domestic animals and has mentioned on Page 35-44 of the proceedings).

83. Thapar GS (1956). Systematic survey of helminth parasites of domesticated animals in India. *Indian Journal of Veterinary Sciences.* 26 : 211-271.

84. Yamaguti S (1958). Systema Helminthum. Vol I. The digenetic trematodes of vertebrates Part I and Part II. Interscience Publishers Inc. New York page 1-938 (for part I), page 981-1575 for part II which contains systematic survey of the digenea of vertebrates and their host relationships (the volume III published by Yamaguti in 1961 is divided in Part I which deals nematodes of fish, amphibians, reptiles, birds and mammals (page 1-680) while Part II deals with systematic survey of nematodes of vertebrates and their host relationships (Page 681-1125).

85. Srivastava HD and Dutt SC 1963. Studies on the life history of *Stephanofilaria assamensis* the causative parasite of hump-sore of Indian cattle. *Indian Journal of Veterinary Sciences* 33 : 173-177

86. This is the site (www.indianpatents.org.in/faqpat.htm) of Patent facilitating centre which helps to answer general questions and how to patent your research work.

87. Kruatrachue M, Bhaibulaya M and Harinasuta C (1965). *Orientobilharzia harnasutai* sp.nov., a mammalian blood fluke,its morphology and life cycle. *Annals of Tropical Medicine and Parasitology* 59:181-188.

88. Illinois Wesleyan University is constructing the Past and in article 8, Buhr Shawn has discussed "To inoculate or not to inoculate ? The debate and the small pox epidemic of Boston in 1721 you may

read the article by logging-digitalcommons.iwu.edu/egi/view content.egi? article=1071 & context=constructing (reference of statistic figures taken from Shyrock, Medicine in America 7). The two persons involved in 'inoculation' experiments were Cotton Mather and Dr Zabdiel Boylston ;As Mather was a minister and American clerical practitioner, medical fraternity, particularly Douglass, strongly opposed "inoculation" procedure (have some idea of controversy by logging en.wikipedia.org/wiki/Cotton_Mather).

89. Lal Chand 1981, Statistics for Beginners. Manojavam Prakashan, Naini, Allahabad Page 172. (or may consult Gupta SC and Kapoor VK 2002. Fundamentals of mathematical statistics. 11th Edition, Sultan Chand and Sons, New Delhi).

90. You can have basic knowledge on camera lucida (a latin word meaning "light chamber") by logging en.wikipedia.org/wiki/Camera-lucida (a book on Camera lucida was published in 1980 by Roland Barthes).

PART D

91. High School Grammar and composition by Wren and Martin, Revised by NDV Prasad in 2010 and published by S.Chand, New Delhi, may have one Pocket Oxford or Cambridge Dictionary, the Webster Dictionary is better as it contains many scientific words and etymology; everyman's thesaurus of English words and phrases to know multiple words with little different meaning; Dorland's Pocket Medical Dictionary; Black's Veterinary Dictionary; now many dictionaries are available on line and on smart phone.

92. Agrawal MC (1974). Studies on the prevalence, morphology, transmission, pathology and chemotherapy of *Stephanofilaria zaheeri* (Singh, 1958), the parasite of ear-sore in buffaloes. MVSc Thesis, Jawaharlal Nehru Krishi Vishwa Vidyalaya, Jabalpur.

93. Kaur Ninder (1985). Comparative studies on serodiagnosis and faecal examination in experimental schistosomiasis. Ph.D thesis, Rani Durgawati Vishwa Vidyalaya, Jabalpur.

94. Gupta Samidha (2002). Clinical,biochemical and parasitological studies and prevalence of caprine schistosomiasis in and around Jabalpur. Ph.D thesis, Rani Durgawati Vishwa Vidyalaya, Jabalpur.

95. Shames (1999). Chemotherapeutic efficacy of praziquantel and closantel in experimental porcine schistosomiasis. MVSc Thesis, Jawaharlal Nehru Krishi Vishwa Vidyalaya, Jabalpur.

96. Smithers SR and Terry RJ (1965). The infection of laboratory hosts with cercariae of *Schistosoma mansoni* and the recovery of the adult worms. *Parasitology* 55: 695-700.

97. Agrawal MC, Sahasrabudhe VK and Gehlot K (1979). Immunization against *schistosoma incognitum* in mice by administration of cercariae of *Schistosoma indicum*. *Indian Veterinary Journal* 56: 682-685.

98. Win T, Dhungyel OP and Khatri G (1991). Parasites of Mithun cattle in Eastern Bhutan. Bhutan Journal of Animal Sciences 12 : 81-84 (for Indian reference, see, Rajkhowa S, Bujarbaruah KM, Rajkhowa C and Thong K 2005. Incidence of intestinal parasitism in Mithun (*Bos frontalis*). *Journal of Veterinary Parasitology* 19: 39-41).

99. Gadgil RK and Shah SN (1952). Human schistosomiasis in India. *Journal of Medical Sciences* 6: 760-763.

100. Sathe BD, Pandit CH, Chandekar NG, Badade DC, Sengupta SR and Renapurkar DM (1991). Serodiagnosis of schistosomiasis by ELISA test in an endemic area of Gimvi Village, India. *Journal of Tropical Medicine and Hygiene*. 94: 76-78.

101. Agrawal MC, Banerjee PS and Shah HL (1991). Five mammalian schistosome species in an endemic focus in India. *Transactions of Royal Society of Tropical Medicine and Hygiene* 85: 231.

Zeitfracht Medien GmbH
Ferdinand-Jühlke-Straße 7
99095 Erfurt, Deutschland
produktsicherheit@kolibri360.de